180 Days of GEOGRAPHY
for Kindergarten

Author

Jessica Hathaway, M.S.Ed.

SHELL EDUCATION

Series Consultant

Nicholas Baker, Ed.D.
Supervisor of Curriculum and Instruction
Colonial School District, DE

Publishing Credits

Corinne Burton, M.A.Ed., *Publisher*
Conni Medina, M.A.Ed., *Managing Editor*
Emily R. Smith, M.A.Ed., *Content Director*
Veronique Bos, *Creative Director*
Shaun N. Bernadou, *Art Director*
Lynette Ordoñez, *Editor*
Jodene Lynn Smith, M.A., *Editor*
Kevin Pham, *Graphic Designer*
Stephanie Bernard, *Associate Editor*

Image Credits

p.112 (bottom right) Martin Good/Shutterstock; p.122 (top) Randrei/Shutterstock;
all other images from iStock and/or Shutterstock.

Standards

© 2012 National Council for Geographic Education
© 2014 Mid-continent Research for Education and Learning (McREL)

For information on how this resource meets national and other state standards,
see pages 10–14. You may also review this information by visiting our website
at www.teachercreatedmaterials.com/administrators/correlations/ and
following the on-screen directions.

Shell Education

A division of Teacher Created Materials
5301 Oceanus Drive
Huntington Beach, CA 92649-1030
www.tcmpub.com/shell-education

ISBN 978-1-4258-3301-5

© 2018 Shell Educational Publishing, Inc.

TABLE OF CONTENTS

INTRODUCTION

With today's geographic technology, the world seems smaller than ever. Satellites can accurately measure the distance between any two points on the planet and give detailed instructions about how to get there in real time. This may lead some people to wonder why we still study geography.

While technology is helpful, it isn't always accurate. We may need to find detours around construction, use a trail map, outsmart our technology, and even be the creators of the next navigational technology.

But geography is also the study of cultures and how people interact with the physical world. People change the environment, and the environment affects how people live. People divide the land for a variety of reasons. Yet no matter how it is divided or why, people are at the heart of these decisions. To be responsible and civically engaged, students must learn to think in geographical terms.

The Need for Practice

To be successful in geography, students must understand how the physical world affects humanity. They must not only master map skills but also learn how to look at the world through a geographical lens. Through repeated practice, students will learn how a variety of factors affect the world in which they live.

Understanding Assessment

In addition to providing opportunities for frequent practice, teachers must be able to assess students' geographical understandings. This allows teachers to adequately address students' misconceptions, build on their current understandings, and challenge them appropriately. Assessment is a long-term process that involves careful analysis of student responses from a discussion, project, practice sheet, or test. The data gathered from assessments should be used to inform instruction: slow down, speed up, or reteach. This type of assessment is called *formative assessment*.

HOW TO USE THIS BOOK

Weekly Structure

The first two weeks of the book focus on map skills. By introducing these skills early in the year, students will have a strong foundation on which to build throughout the year. Each of the remaining 34 weeks will follow a regular weekly structure.

Each week, students will study a grade-level geography topic and a location in a community. Locations may be a town, a state, a region, or the whole continent.

Days 1 and 2 of each week focus on map skills. Days 3 and 4 allow students to apply information and data to what they have learned. Day 5 helps students connect what they have learned to themselves.

 Day 1—Reading Maps: Students will study a grade-appropriate map and answer questions about it.

 Day 2—Creating Maps: Students will create maps or add to an existing map.

 Day 3—Read About It: Students will read a text related to the topic or location for the week and answer text-dependent or photo-dependent questions about it.

 Day 4—Think About It: Students will analyze a chart, diagram, or other graphic related to the topic or location for the week and answer questions about it.

 Day 5—Geography and Me: Students will do an activity to connect what they learned to themselves.

Five Themes of Geography

Good geography teaching encompasses all five themes of geography: location, place, human-environment interaction, movement, and region. Location refers to physical and absolute and relative locations or a specific point or place. The place theme refers to the human characteristics of a place. Human-environment interaction describes how humans affect their surroundings and how the environment affects the people who live there. Movement describes how and why people, goods, and ideas move between different places. The region theme examines how places are grouped into different regions. Regions can be divided based on a variety of factors, including physical characteristics, cultures, weather, political factors, and many others.

HOW TO USE THIS BOOK *(cont.)*

Weekly Themes

The following chart shows the topics, locations, and themes of geography that are covered during each week of instruction.

Wk.	Topic	Location	Geography Themes
1	—Map Skills Only—		Location
2			Location
3	Over/Under	Kitchen	Location
4	Near/Far	Playground	Location
5	Left/Right	Library	Location
6	Behind/In Front	Garden	Location
7	Relative Location	Bedroom	Location
8	Address (absolute location)	Neighborhood	Location
9	Land and Water	World	Location, Place
10	Landforms	Wilderness	Place
11	Bodies of Water	Wilderness	Place
12	Weather	Neighborhood	Location, Place
13	Climate	Desert	Place
14	Vegetation	Wilderness	Place, Region
15	Ecosystems	Pond	Place
16	Clothing	Landscape	Place, Human-Environment Interaction
17	Activities/Weather	Lake/Mountain	Place, Human-Environment Interaction

HOW TO USE THIS BOOK *(cont.)*

Wk.	Topic	Location	Geography Themes
18	Types of Shelter	Landscape	Human-Environment Interaction
19	Homes Over Time	Community	Place
20	Types of Settlements	Community	Place
21	Moving	Farm/City	Movement
22	Land Development	Community	Place, Human-Environment Interaction
23	Community Helpers	Town	Place
24	Community	School	Location, Place
25	Demographics	Community	Place
26	Jobs	Town	Place, Human-Environment Interaction
27	Goods and Services	Town	Place
28	Types of Transportation	City	Movement
29	Transporting Goods	Community	Movement
30	Natural Resources	Community	Human-Environment Interaction
31	Resource Management	Community	Human-Environment Interaction
32	Family	Home	Place
33	Food	Community	Place
34	Language	World	Place
35	Cooperation	Community	Human-Environment Interaction
36	Disputes and Resolutions	Community	Place

HOW TO USE THIS BOOK *(cont.)*

Using the Practice Pages

The activity pages provide practice and assessment opportunities for each day of the school year. Teachers may wish to prepare packets of weekly practice pages for the classroom or for homework.

As outlined on page 4, each week examines one location and one geography topic.

 The first two days focus on map skills. On Day 1, students will study a map and answer questions about it. On Day 2, they will add to or create a map.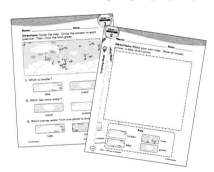

 Days 3 and 4 allow students to apply information and data from texts, charts, graphs, and other sources to the location being studied.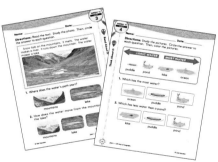

 On Day 5, students will apply what they learned to themselves.

Using the Resources

Rubrics for the types of days (map skills, applying information and data, and making connections) can be found on pages 202–204 and in the Digital Resources. Use the rubrics to assess students' work. Be sure to share these rubrics with students often so that they know what is expected of them.

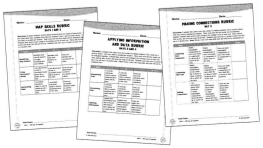

HOW TO USE THIS BOOK *(cont.)*

Diagnostic Assessment

Teachers can use the practice pages as diagnostic assessments. The data analysis tools included with the book enable teachers or parents to quickly score students' work and monitor their progress. Teachers and parents can quickly see which skills students may need to target further to develop proficiency.

Students will learn map skills, how to apply text and data to what they have learned, and how to relate what they learned to themselves. Teachers can assess students' learning in each area using the rubrics on pages 202–204. Then, record their scores on the Practice Page Item Analysis sheets on pages 205–207. These charts are also provided in the Digital Resources as PDFs, Microsoft Word® files, and Microsoft Excel® files (see page 208 for more information). Teachers can input data into the electronic files directly on the computer, or they can print the pages.

To Complete the Practice Page Item Analyses:

- Write or type students' names in the far-left column. Depending on the number of students, more than one copy of the forms may be needed.

  - The skills are indicated across the tops of the pages.

  - The weeks in which students should be assessed are indicated in the first rows of the charts. Students should be assessed at the ends of those weeks.

- Review students' work for the days indicated in the chart. For example, if using the Making Connections Analysis sheet for the first time, review students' work from Day 5 for all five weeks.

- Add the scores for each student. Place that sum in the far right column. Record the class average in the last row. Use these scores as benchmarks to determine how students are performing.

Digital Resources

The Digital Resources contain digital copies of the rubrics, analysis pages, and standards charts. See page 208 for more information.

HOW TO USE THIS BOOK *(cont.)*

Using the Results to Differentiate Instruction

Once results are gathered and analyzed, teachers can use them to inform the way they differentiate instruction. The data can help determine which geography skills are the most difficult for students and which students need additional instructional support and continued practice.

Whole-Class Support

The results of the diagnostic analysis may show that the entire class is struggling with certain geography skills. If these concepts have been taught in the past, this indicates that further instruction or reteaching is necessary. If these concepts have not been taught in the past, this data is a great preassessment and may demonstrate that students do not have a working knowledge of the concepts. Thus, careful planning for the length of the unit(s) or lesson(s) must be considered, and additional front-loading may be required.

Small-Group or Individual Support

The results of the diagnostic analysis may show that an individual student or a small group of students is struggling with certain geography skills. If these concepts have been taught in the past, this indicates that further instruction or reteaching is necessary. Consider pulling these students aside to instruct them further on the concepts while others are working independently. Students may also benefit from extra practice using games or computer-based resources.

Teachers can also use the results to help identify proficient individual students or groups of students who are ready for enrichment or above-grade-level instruction. These students may benefit from independent learning contracts or more challenging activities.

STANDARDS CORRELATIONS

Shell Education is committed to producing educational materials that are research and standards based. In this effort, we have correlated all our products to the academic standards of all 50 states, the District of Columbia, the Department of Defense Dependents Schools, and all Canadian provinces.

How to Find Standards Correlations

To print a customized correlation report of this product for your state, visit our website at **www.teachercreatedmaterials.com/administrators/correlations** and follow the on-screen directions. If you require assistance in printing correlation reports, please contact our Customer Service Department at 1-877-777-3450.

Purpose and Intent of Standards

The Every Student Succeeds Act (ESSA) mandates that all states adopt challenging academic standards that help students meet the goal of college and career readiness. While many states already adopted academic standards prior to ESSA, the act continues to hold states accountable for detailed and comprehensive standards. Standards are designed to focus instruction and guide adoption of curricula. Standards are statements that describe the criteria necessary for students to meet specific academic goals. They define the knowledge, skills, and content students should acquire at each level. Standards are also used to develop standardized tests to evaluate students' academic progress. Teachers are required to demonstrate how their lessons meet state standards. State standards are used in the development of our products, so educators can be assured they meet the academic requirements of each state.

The activities in this book are aligned to the National Geography Standards and the McREL standards. The chart on pages 11–12 lists the National Geography Standards used throughout this book. The chart on pages 13–14 correlates the specific McREL and National Geography Standards to each week. The standards charts are also in the Digital Resources (standards.pdf).

C3 Framework

This book also correlates to the College, Career, and Civic Life (C3) Framework published by the National Council for the Social Studies. By completing the activities in this book, students will learn to answer and develop strong questions (Dimension 1), critically think like a geographer (Dimension 2), and effectively choose and use geography resources (Dimension 3). Many activities also encourage students to take informed action within their communities (Dimension 4).

STANDARDS CORRELATIONS *(cont.)*

180 Days of Geography is designed to give students daily practice in geography through engaging activities. Students will learn map skills, how to apply information and data to their understandings of various locations and cultures, and how to apply what they learned to themselves.

Easy to Use and Standards Based

There are 18 National Geography Standards, which fall under six essential elements. Specific expectations are given for fourth grade, eighth grade, and twelfth grade. For this book, fourth grade expectations were used with the understanding that full mastery is not expected until that grade level.

Essential Elements	National Geography Standards
The World in Spatial Terms	**Standard 1:** How to use maps and other geographic representations, geospatial technologies, and spatial thinking to understand and communicate information
	Standard 2: How to use mental maps to organize information about people, places, and environments in a spatial context
	Standard 3: How to analyze the spatial organization of people, places, and environments on Earth's surface
Places and Regions	**Standard 4:** The physical and human characteristics of places
	Standard 5: People create regions to interpret Earth's complexity
	Standard 6: How culture and experience influence people's perceptions of places and regions
Physical Systems	**Standard 7:** The physical processes that shape the patterns of Earth's surface
	Standard 8: The characteristics and spatial distribution of ecosystems and biomes on Earth's surface

STANDARDS CORRELATIONS *(cont.)*

Essential Elements	National Geography Standards
Human Systems	**Standard 9:** The characteristics, distribution, and migration of human populations on Earth's surface
	Standard 10: The characteristics, distribution, and complexity of Earth's cultural mosaics
	Standard 11: The patterns and networks of economic interdependence on Earth's surface
	Standard 12: The process, patterns, and functions of human settlement
	Standard 13: How the forces of cooperation and conflict among people influence the division and control of Earth's surface
Environment and Society	**Standard 14:** How human actions modify the physical environment
	Standard 15: How physical systems affect human systems
	Standard 16: The changes that occur in the meaning, use, distribution, and importance of resources
The Uses of Geography	**Standard 17:** How to apply geography to interpret the past
	Standard 18: How to apply geography to interpret the present and plan for the future

–2012 National Council for Geographic Education

STANDARDS CORRELATIONS *(cont.)*

Easy to Use and Standards Based *(cont.)*

This chart lists the specific National Geography Standards (NGS) and McREL standards that are covered each week.

Wk.	NGS	McREL Standards
1	Standards 1 and 3	Uses simple geographic thinking.
2	Standards 1 and 3	Understands the globe as a representation of the Earth. Uses simple geographic thinking.
3	Standard 3	Identifies physical and human features in terms of the four spatial elements.
4	Standard 3	Identifies physical and human features in terms of the four spatial elements.
5	Standard 3	Identifies physical and human features in terms of the four spatial elements.
6	Standard 3	Identifies physical and human features in terms of the four spatial elements.
7	Standard 3	Identifies physical and human features in terms of the four spatial elements.
8	Standard 1	Knows the location of school, home, neighborhood, community, state, and country.
9	Standards 3 and 7	Knows that places can be defined in terms of their predominant human and physical characteristics.
10	Standards 3 and 7	Knows that places can be defined in terms of their predominant human and physical characteristics.
11	Standards 3 and 7	Knows that places can be defined in terms of their predominant human and physical characteristics.
12	Standard 7	Knows that places can be defined in terms of their predominant human and physical characteristics.
13	Standard 8	Knows that places can be defined in terms of their predominant human and physical characteristics.
14	Standard 8	Knows areas that can be classified as regions according to physical criteria.
15	Standard 8	Knows that places can be defined in terms of their predominant human and physical characteristics.
16	Standard 15	Knows that places can be defined in terms of their predominant human and physical characteristics.
17	Standard 15	Knows that places can be defined in terms of their predominant human and physical characteristics.

STANDARDS CORRELATIONS *(cont.)*

Wk.	NGS	McREL Standards
18	Standards 9 and 15	Knows the similarities and differences in housing and land use in urban and suburban areas.
19	Standard 9	Knows how areas of a community have changed over time.
20	Standard 12	Knows the similarities and differences in housing and land use in urban and suburban areas.
21	Standards 9 and 12	Understands why people choose to settle in different places.
22	Standards 12 and 17	Knows how areas of a community have changed over time.
23	Standard 12	Knows that places can be defined in terms of their predominant human and physical characteristics.
24	Standard 4	Knows that places can be defined in terms of their predominant human and physical characteristics.
25	Standard 9	Knows that places can be defined in terms of their predominant human and physical characteristics.
26	Standard 11	Knows that places can be defined in terms of their predominant human and physical characteristics.
27	Standard 11	Knows the physical and human characteristics of the local community.
28	Standard 11	Knows the modes of transportation used to move people, products and ideas from place to place, their importance and their advantages and disadvantages.
29	Standard 11	Knows the modes of transportation used to move people, products and ideas from place to place, their importance and their advantages and disadvantages.
30	Standard 16	Knows ways in which people depend on the physical environment. Knows the role that resources play in our daily lives.
31	Standard 16	Knows ways in which people depend on the physical environment.
32	Standard 10	Knows the basic components of culture.
33	Standard 10	Knows the basic components of culture.
34	Standard 10	Knows the basic components of culture.
35	Standard 13	Knows ways that people solve common problems by cooperating.
36	Standard 13	Knows ways that people solve common problems by cooperating.

Name: _____ **Date:** _____

Directions: Study the map. Follow the steps.

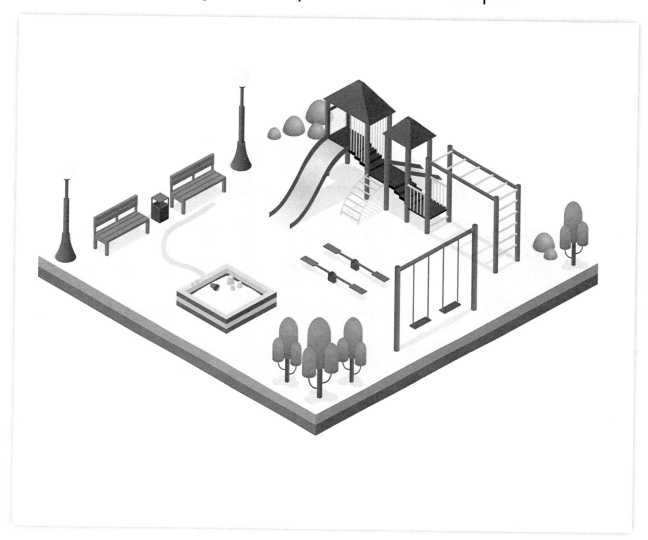

1. Draw a person near the slide.

2. Circle something that is far from the sandbox.

3. Draw a box on the map.

4. Tell a friend what your box is near.

Read About It

Name: _____ **Date:** _____

Directions: Read the text. Study the photo. Then, answer the questions.

Come play in the sand! We can dig. Look at our hole! We can build. Look at our house! Playing with sand is fun.

1. Find the boy with the striped shirt. Draw a bucket near him.

2. Is the girl near or far from the boys? Circle the answer.

   near   far

3. Circle the flowers that are far from the kids.

Name: _____ **Date:** _____

Directions: Study the graph. Circle the answer to each question.

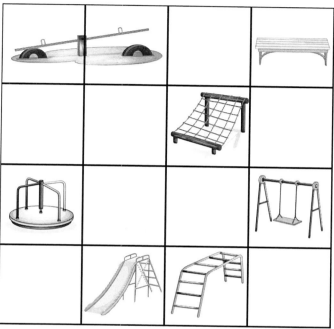

1. How far is the slide from the swing?

   1 square   3 squares   5 squares

2. How far is the swing from the bench?

   2 squares   4 squares   8 squares

3. Which is nearer to the monkey bars?

   slide   bench

Geography and Me

Name: _____ **Date:** _____

Directions: Draw a playground. Draw yourself near the playground. Draw something far from the playground.

Name: _____ **Date:** _____

Directions: Study the picture. Circle the answer to each question.

1. What side of the library has animal books?

left   right

2. What books are on the right side of the library?

animals   sports   food

3. The ladder is on the _____ side of the library.

left   right

Name: _____   **Date:** _____

Creating Maps

Directions: Study the picture. Follow the steps.

1. Draw a circle on the right side of the desk.

2. Put an X on the ladder.

3. Draw yourself to the right of a pet.

Name: _____ **Date:** _____

Directions: Read the text. Study the photo. Then, circle the answer to each question.

This is a library. It has books. People borrow the books. They use a library card. They take the books home. They read them. Then, they return the books.

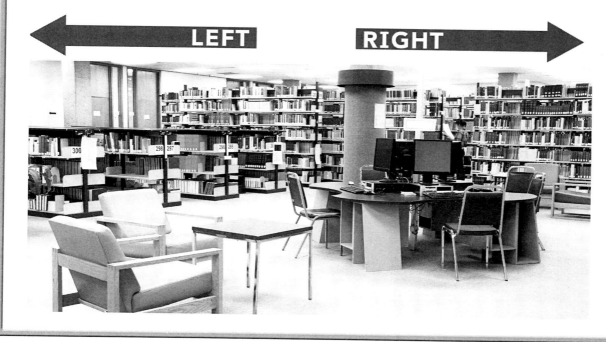

1. What is on the left side of the library?

computer   chair   backpack

2. Which side of the photo is the computer on?

left   right

Think About It

Name: _____ **Date:** _____

Directions: Follow the steps to draw Maria's path.

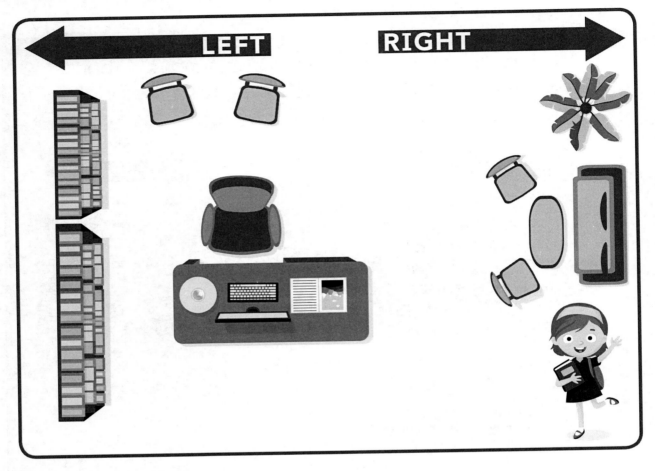

1. Maria walked straight to the desk.

2. Then, she turned and walked to the plant.

3. Then, she turned around and walked straight. She walked to the books.

Try It! Tell a friend what is on the left and right side of the library.

© Shell Education

Name: _____ **Date:** _____

Directions: Look at your classroom. What is on the left side of the room? What is on the right side? Draw the objects in the chart.

Left	Right

Reading Maps

Name: _____ **Date:** _____

Directions: Study the picture. Circle the answer to each question. Then, color the picture.

1. What is in front of the plants?

building   rabbit   gardener

2. What is behind the gardener?

plants   child   turtle

3. The buildings are _____ the garden.

in front of   behind

Name: _____ **Date:** _____

Directions: Study the picture. Follow the steps.

1. Draw an animal in front of the gardener.

2. Circle a plant behind the gardener.

Try It! Tell a friend what is in front of you.

Read About It

Name: _____ **Date:** _____

Directions: Read the text. Study the picture. Circle the answer to each question. Then, color the picture.

Jack is a rabbit. He saw a fox. The fox chased Jack. Jack hid behind a bush. The fox went away. Then, Jack went to the garden. He ate a tasty lunch. Jack loves vegetables!

1. What is the fox sitting in front of?

   tree   bush   fox

2. What did Jack hide behind?

   tree   corn   bush

Name: _____ **Date:** _____

Directions: Study the picture. Circle the answer to each question.

Key
carrot
lettuce
radish
mint

1. What plant is behind the radishes?

corn   mint   lettuce

2. What plant is in front of the radishes?

mint   carrots   lettuce

3. Are the carrots in front of or behind the lettuce?

in front   behind

Name: _____ **Date:** _____

Geography and Me

Directions: Color the person to look like you. Draw your favorite fruits in front of you. Draw your least favorite fruits behind you.

Name: _____ **Date:** _____

Directions: Study the picture. Circle the answer to each question. Then, color the picture.

1. What is over the bed?

map   clock   skateboard

2. Is the lamp on the left or right side of the bedroom?

left   right

3. What is in front of the bookcase?

bed   lamp   skateboard

Name: _____ **Date:** _____

Directions: Study the picture. Follow the steps.

Creating Maps

1. Circle the object that is above the bed.

2. Draw a toy on the left side of the room.

3. Draw a cat on the right side of the room.

4. Put an X on something near the bed.

5. Draw a star far from the lamp.

> **Try It!** Tell a friend what is near the lamp. Tell what is far from the lamp.

Name: _____ **Date:** _____

Directions: Read the text. Study the photo. Then, circle the answer to each question.

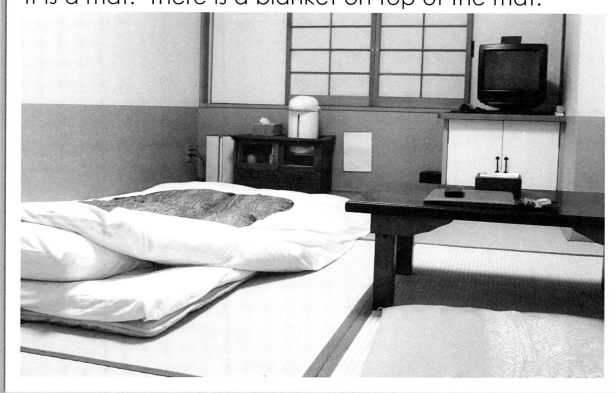

This is a bedroom in Japan. The bed is on the floor. It is a mat. There is a blanket on top of the mat.

1. What is under the blanket?

chair   mat   TV

2. The bed is on the _____ side of the room.

left   right

Name: _____ **Date:** _____

Directions: Jeff's blocks are on the floor. Study the chart. Draw the blocks in the picture.

Position	Number of Blocks
in front of box	3
behind box	1
left side of box	2
right side of box	4

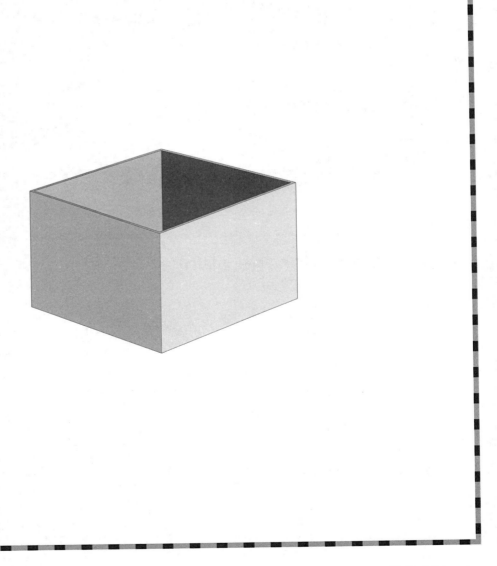

Name: _____ **Date:** _____

Directions: Look around you. What do you see? Draw objects to complete the table.

Position	Object
in front of you	
over you	
on your left	
near you	

Geography and Me

Reading Maps

Name: _____ **Date:** _____

Directions: Study the map. Circle the answer to each question. Then, color the map.

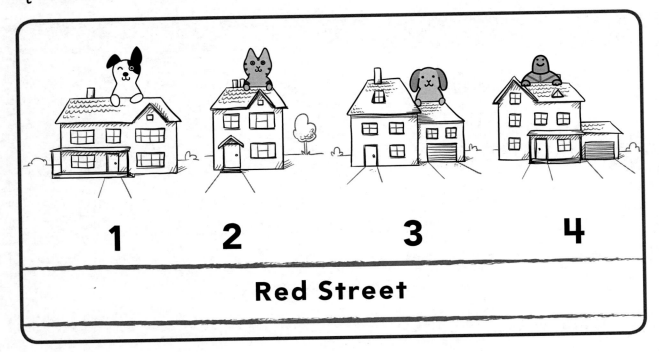

Red Street

1. What is the name of the street?

Green Street   Blue Street   Red Street

2. Who lives in House 2?

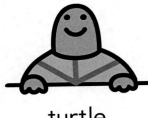      

turtle   rabbit   cat

3. Which house does the turtle live in?

1   2   3   4

Name: _____ **Date:** _____

Directions: Study the map. Follow the steps.

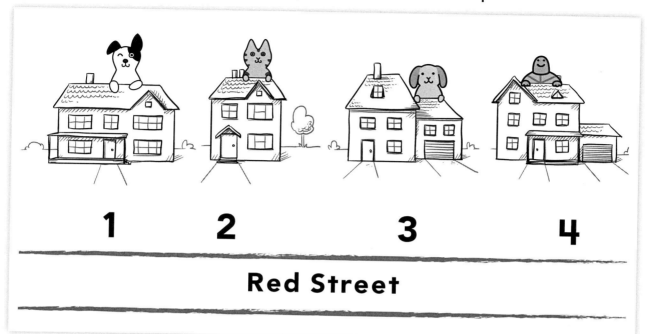

1   2   3   4

Red Street

1. Color the street red.

2. Put an X on House 4.

3. Draw a path to get from the turtle's house to the cat's house.

4. Draw where you live.

Read About It

Name: _____ **Date:** _____

Directions: Read the text. Study the picture. Then, circle the answer to each question.

The Jones family is moving. They pack a truck. They drive to the new house. They look for the address. It helps them find the house. They unpack. They love their new house!

Park Street

1. Circle the name of the Jones's new street.

Park Street   Oak Street   Main Street

2. Circle the number of the Jones's new house.

3   1   8

Name: _____ **Date:** _____

Directions: Study the envelope. Circle the answer to each question.

1. What is the house number?

Main Street   123   CA

2. What is the town?

Main Street   54321   Avalon

3. What is the name?

The Long Family   Avalon   123

Think About It

Name: _____ **Date:** _____

Directions: Write your address on the envelope. Then, draw your home.

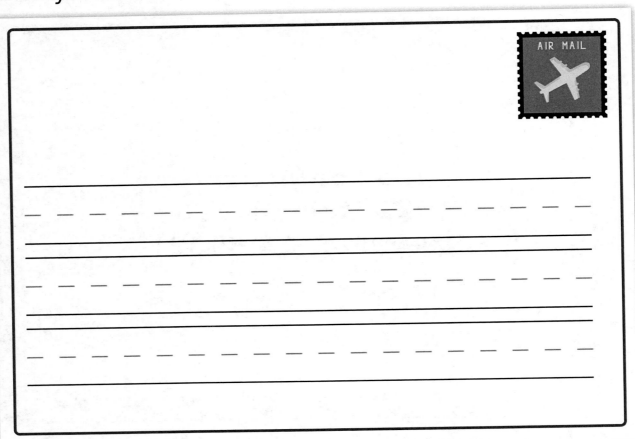

Name: _____ **Date:** _____

Directions: Study the map. Answer the questions.

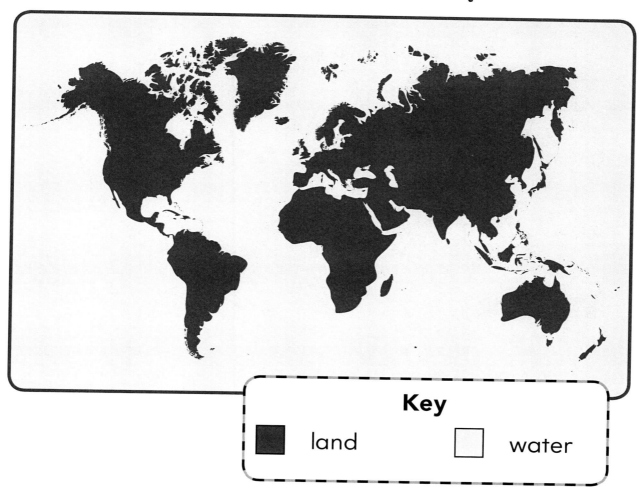

Key

land ◼   water ☐

1. Circle the color that shows land.

2. Circle the color that shows water.

3. Tell a friend how to use the key.

Creating Maps

Name: _____ **Date:** _____

Directions: Study the map. Follow the steps.

1. Color the land green.

2. Color the water blue.

3. Draw a place that has land and water.

Name: _____ **Date:** _____

Directions: Read the text. Study the photo. Then, answer the questions.

Tampa is a city. It is near the water. Tampa has beaches. It has homes. People live there. They work there, too.

1. What do you find where the land meets the water in Tampa? Circle your answer.

beach

forest

mountain

2. How do people in Tampa use the water nearby? Draw your answer.

Name: _____ **Date:** _____

Directions: Study the map. Follow the steps.

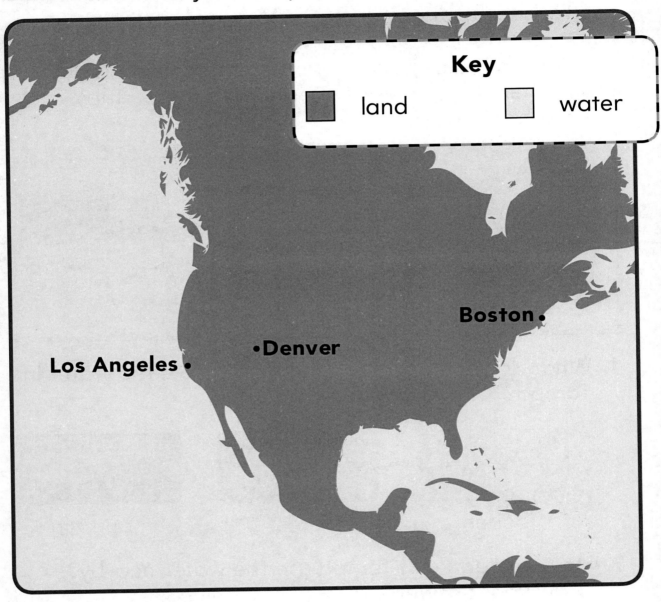

Key

□ land   □ water

Boston •

•Denver

Los Angeles •

1. Circle the cities that are near the water.

2. Put an X on the city that is not near the water.

3. Which city is closer to the water? Circle your answer.

Los Angeles   Denver

Name: _____ **Date:** _____

Directions: Draw what you like to do on land. Draw what you like to do in water.

land   water

land	water

Geography and Me

Reading Maps

Name: _____ **Date:** _____

Directions: Study the pictures. Circle the answer to each question. Then, color the pictures.

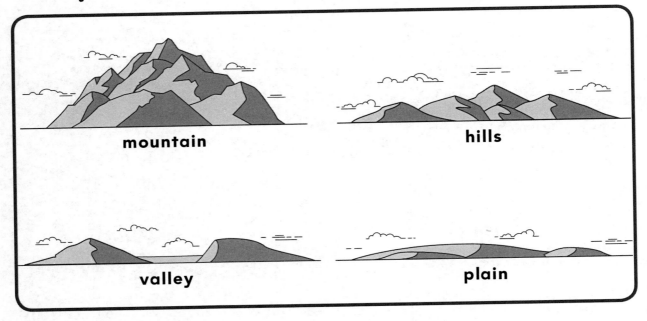

1. Which landform is the tallest?

plain   mountain   valley

2. Which landform is very flat?

plain   hills   mountain

3. Tell a friend how each landform is different.

Name: _____ **Date:** _____

Directions: Draw the landforms.

mountain

hill

valley

plain

Creating Maps

Read About It

Name: _____ **Date:** _____

Directions: Read the text. Study the photo. Then, follow the steps.

The land has many forms. Some places have tall mountains. Some places have flat plains. A valley is a low area. It is between two hills or mountains. People live on all types of land.

1. Draw a tree at the top of the mountain.

2. A valley is between two _____.

hills

plains

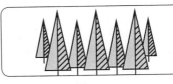

trees

Name: _____ **Date:** _____

Directions: This map shows how the bear got to his cave. Study the map. Circle the answer to each question. Then, color the map.

1. Where did the bear start his journey?

    plain   valley   mountain

2. Where is the bear's cave?

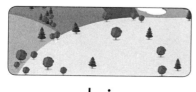 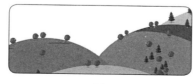 

    plain   valley   mountain

Geography and Me

Name: _____ **Date:** _____

Directions: Circle the landforms that are near your home. Then, draw a picture of the land near your home.

plain

valley

hills

mountains

Name: _____ **Date:** _____

Directions: Study the map. Circle the answer to each question. Then, color the land green.

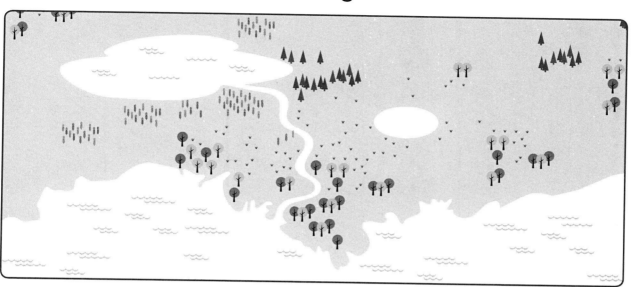

1. Which is smaller?

lake

pond

2. Which has more water?

pond

ocean

3. Which carries water from one place to another?

lake

river

Creating Maps

Name: _____ **Date:** _____

Directions: Make your own map. Draw an ocean, a river, a lake, and a pond.

Key

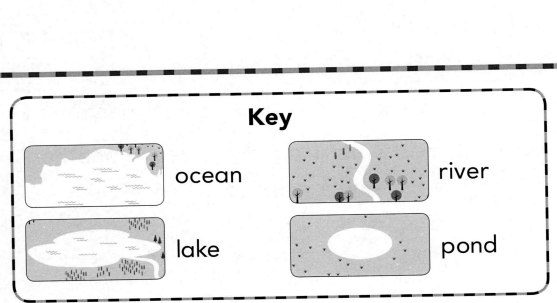

Name: _____ **Date:** _____

Directions: Read the text. Study the photo. Then, circle the answer to each question.

> Snow falls on the mountains. It melts. The water makes a river. It runs down the mountain. The water makes a lake.

1. Where does the water's path start?

mountains

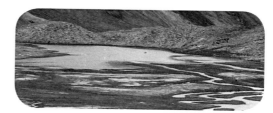

lake

2. How does the water move from the mountain to the lake?

mountains

lake

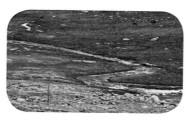

river

Think About It

Name: _____ **Date:** _____

Directions: Study the pictures. Circle the answer to each question. Then, color the pictures.

LEAST WATER   MOST WATER

puddle   pond   lake   ocean

1. Which has the most water?

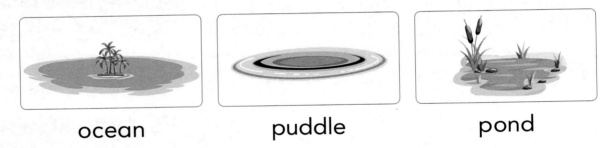

ocean   puddle   pond

2. Which has less water than a pond?

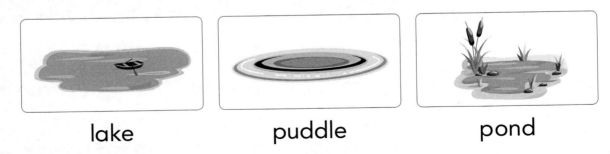

lake   puddle   pond

Name: _____ **Date:** _____

Directions: Draw a picture of a time you visited the water.

Reading Maps

Name: _____ **Date:** _____

Directions: Study the map. Circle the answer to each question. Then, color the map.

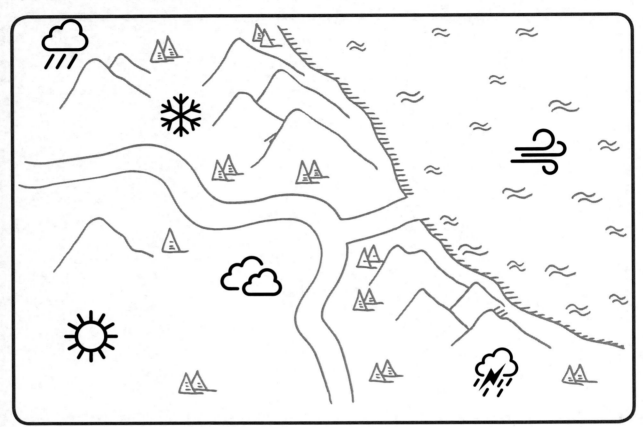

1. Which type of weather is over the ocean?

sunny　　　　　　snowy　　　　　　windy

2. Which type of weather gets you wet?

sunny　　　　　　windy　　　　　　rainy

Name: _____ **Date:** _____

Directions: Draw a different type of weather at each house. Label the weather on the line.

Word Bank

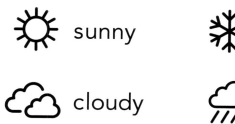 sunny

 snowy

 windy

 cloudy

 rainy

 stormy

Creating Maps

Name: _____ **Date:** _____

Read About It

Directions: Read the text. Study the photo. Then, circle the answer to each question.

Storms have many types of weather. Some storms have strong winds. Some have rain or snow. Some storms even have lightning. Storms can be fun to watch.

1. What type of weather do you see in the photo?

rainy

sunny

snowy

2. What are two types of weather storms can have?

rainy

snowy

sunny

Name: _____ **Date:** _____

Directions: Study the chart. Circle the answer to each question. Then, color the symbols.

Winter	Spring	Summer	Fall
❄	☁	☀	☀
🌧	☀		☁
☁	🌧		🌧

1. What is the weather like in summer?

    snowy   rainy   sunny

2. When does it snow?

   winter   spring   summer   fall

3. Does it rain in the spring?

           yes   no

Think About It

Name: _____ **Date:** _____

Directions: Draw weather that you like. Draw weather that you do not like. Use the Word Bank to help you.

Weather I Like	Weather I Do Not Like

Word Bank

 sunny   snowy   windy

cloudy   rainy   stormy

Name: _____ **Date:** _____

Directions: Study the map. Circle the answer to each question.

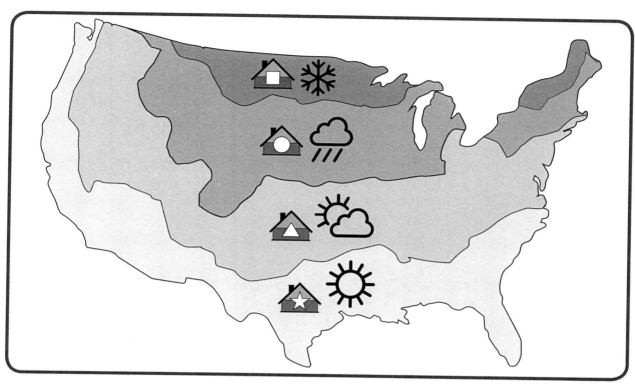

1. Which house is in the coldest climate?

2. Which house is in the hottest climate?

3. Which house is in an area where it rains a lot?

Creating Maps

Name: _____ **Date:** _____

Directions: Use the key to color the map.

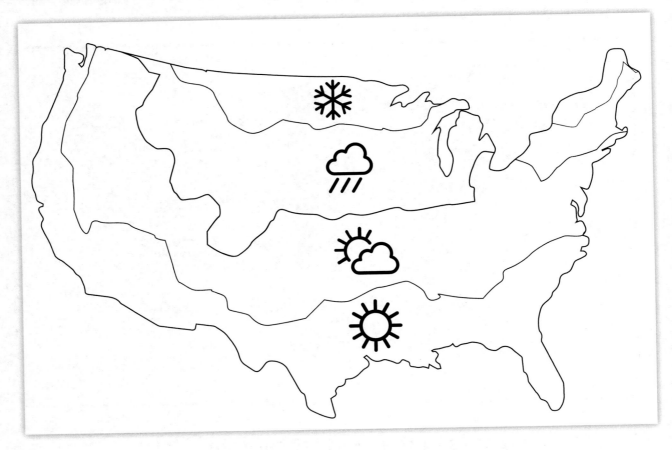

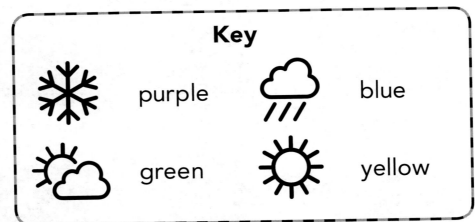

Name: _____ **Date:** _____

Directions: Read the text. Study the picture. Then, answer the questions.

Deserts are hot and dry. Cacti are a type of plant. They live in the desert. They do not need much water. The desert is home to animals, too. Snakes and lizards like hot weather. Mice and foxes live there, too.

1. What is the weather like in the desert?

sunny

rainy

snowy

2. Circle the desert animals in the picture.

3. Cacti have different shapes. Draw two different types of cacti.

Think About It

Name: _____ **Date:** _____

Directions: Study the graph. Answer the questions.

Maya   Manny   Jason   Kate

1. Who lives in the wettest climate? Circle the answer.

Maya   Manny   Jason   Kate

2. How much rain falls at Jason's house?

3. Who lives in the driest climate? Circle the answer.

Maya   Manny   Jason   Kate

Name: _____ **Date:** _____

Directions: Write and draw about the climate where you live. Then, write and draw about the desert.

Where I Live

Desert

Name: _____ **Date:** _____

Directions: Study the map. Circle the answer to each question. Then, color the map.

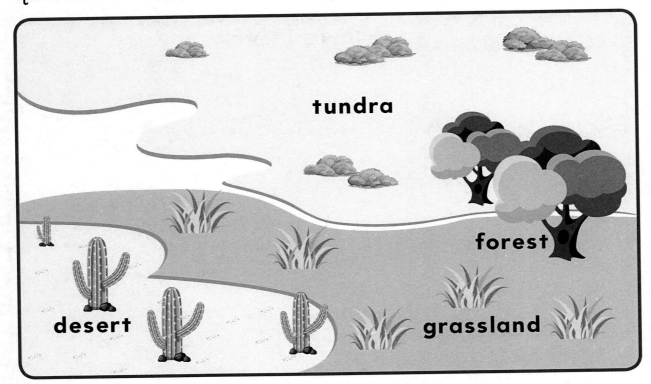

1. What plants grow in a forest?

grass

trees

flowers

2. What plants grow in the desert?

cacti

grass

trees

Name: _____ **Date:** _____

Directions: Follow the steps. Trace the path the person took.

1. The person started in the grassland. Color the grass green.

2. Next, he went to the desert. Circle the desert.

3. Finally, he went to the forest. Draw more trees in the forest.

Try It! Tell a friend about the person's path.

Read About It

Name: _____ **Date:** _____

Directions: Read the text. Study the photo. Then, circle the answer to each question.

The tundra is cold. It is often covered in snow. Only small plants and grass grow there. It is too cold for trees. Reindeer live in the tundra. Polar bears do, too.

1. What type of plants grow in the tundra?

trees   cacti   grass

2. It is too _____ for trees to grow in the tundra.

cold   hot   wet

3. What is the weather like in the tundra?

sunny   snowy   stormy

28621—180 Days of Geography   © *Shell Education*

Name: _____ **Date:** _____

Directions: Study the picture. Circle the answer to each question.

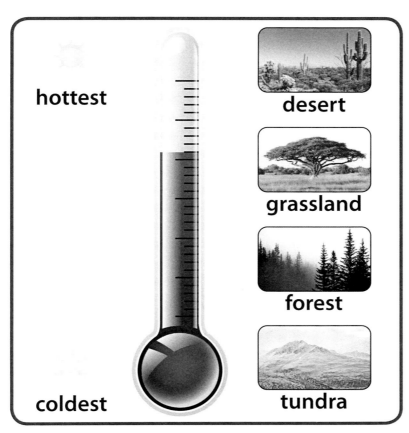

1. Which is the coldest region?

    tundra   forest   desert

2. The _____ is the hottest region.

    tundra   grassland   desert

Name: _____ **Date:** _____

Directions: Take a walk around your school. Draw the plants you see.

Name: _____ **Date:** _____

Directions: Study the picture. Answer the questions.

1. Draw animals that live in or near a pond.

2. Draw plants that live in or near a pond.

Creating Maps

Name: _____ **Date:** _____

Directions: Draw your own map of a pond. Add at least two plants and two animals to your map.

Word Bank

tree · lily pad · bird

fish · turtle · frog

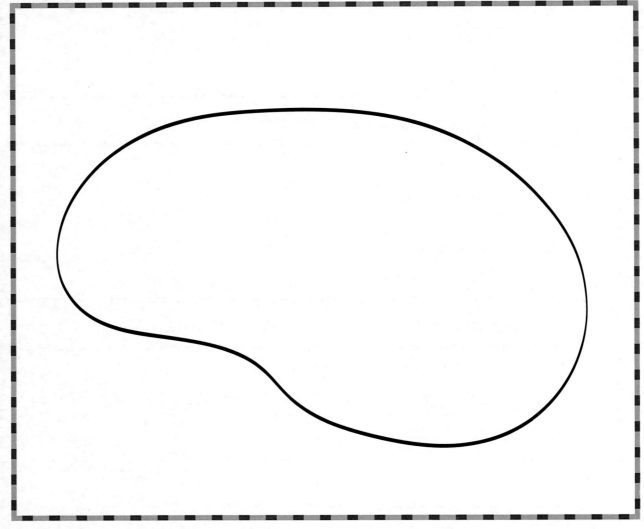

Name: _____ **Date:** _____

Directions: Read the text. Study the photo. Then, circle the answer to each question.

A pond is an ecosystem. Plants and animals live there. This frog sits on a lily pad. It looks for bugs to eat. Then, a bird may eat the frog. Life goes on at the pond.

1. Which would a frog eat?

flower

carrot

bug

2. What pond animal eats frogs?

bird

turtle

fish

3. Which type of plant is in the photo?

tree

bush

lily pad

Think About It

Name: _____ **Date:** _____

Directions: Study the picture. Circle the answer to each question. Then, color the pictures.

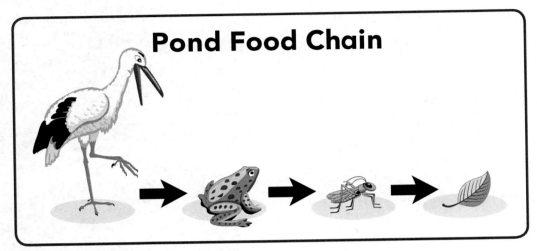

Pond Food Chain

1. What does the grasshopper eat?

frog   leaf   bird

2. Who eats the frog?

grasshopper   leaf   bird

3. What does a frog eat?

grasshopper   leaf   bird

Name: _____ **Date:** _____

Directions: Look at a patch of grass or soil outside. Draw the plants and animals you find.

Reading Maps

Name: _____ **Date:** _____

Directions: Study the map. Circle the answer to each question. Then, color the map.

1. In which type of weather do you need an umbrella?

snowy

rainy

sunny

2. What should you wear when it is snowing?

coat

bathing suit

t-shirt

3. In which type of weather can you build a snowman?

snowy

rainy

sunny

Name: _____ **Date:** _____

Directions: Draw the weather that matches how each person is dressed.

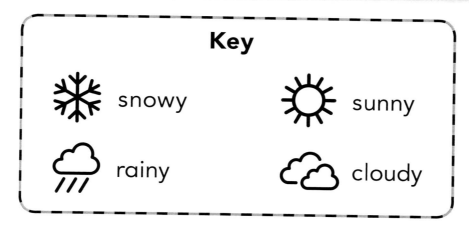

Key

❄ snowy ☀ sunny

🌧 rainy ☁ cloudy

Read About It

Name: _____ **Date:** _____

Directions: Read the text. Study the photo. Circle the answer to each question.

Clothes are important. They protect us. Clothes shade our skin from the sun. Clothes also keep us warm. It is very cold where this woman lives. She wears heavy clothes to stay warm.

1. This woman lives in _____ weather.

        hot   warm   cold

2. What kind of clothes keeps her warm?

    coat   igloo   t-shirt

3. Draw clothes that keep you warm.

Name: _____ **Date:** _____

Directions: Study the picture. Circle the answer to each question.

1. What protects the boy's eyes?

jacket

sunglasses

jeans

2. The shoes protect the boy's _____.

head

hands

feet

3. What kind of weather is the boy dressed for?

snowy

sunny

rainy

Name: _____ **Date:** _____

Directions: Draw the clothes you wear in cold weather. Then, draw the clothes you wear in hot weather.

Cold Weather

Hot Weather

Name: _____ **Date:** _____

Directions: Study the picture. Circle the answer to each question. Then, color the picture.

1. Where can you go sledding?

hill

pond

2. What can you do on a frozen pond?

ski

sled

skate

Creating Maps

Name: _____ **Date:** _____

Directions: Draw lines to show where you can do each summer activity. Tell a friend why you chose each place.

Name: _____ **Date:** _____

Directions: Read the text. Study the photo. Circle the answer to each question.

Planning a vacation is fun! Where will you go? Some places are hot. Hot weather is good for swimming. Some places are cold. Sledding is fun in cold weather. Have fun!

1. Look at the photo. Where did this family go on vacation?

camping   beach   city

2. What kind of weather does this family prefer?

sunny   snowy   rainy

Think About It

Name: _____ Date: _____

Directions: Study the chart. Circle the answer to each question.

	Hot Weather ☀	Cold Weather ❄
mountains		
lake		
field		

1. What can you do at a lake in hot weather?

skate

soccer

swim

2. Where is a good place to ride a bike?

mountain

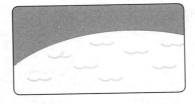
lake

field

Name: _____ **Date:** _____

Directions: Draw where you want to go on vacation. Show what you want to do. Then, write or draw the best weather for your vacation.

Weather: _____

Geography and Me

Name: _____ **Date:** _____

Directions: Study the map. Circle the answer to each question. Then, color the map.

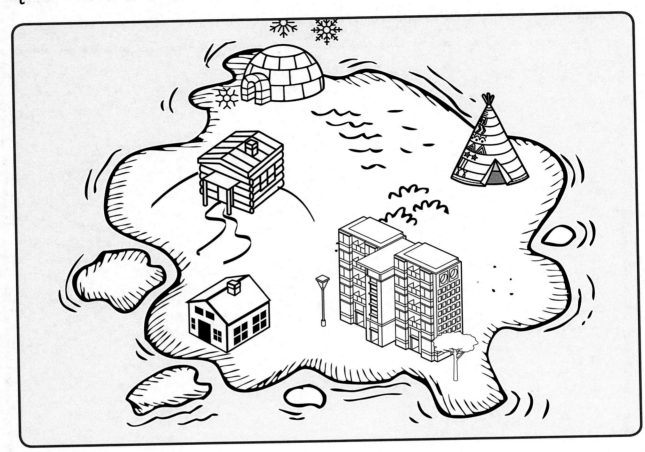

1. Which home is in the snow?

igloo   tepee   house

2. Which home is the tallest?

tepee   apartment   log cabin

Name: _____ **Date:** _____

Directions: Study the map. Follow the steps.

1. Circle the tepee.

2. Draw a person by the apartment building.

3. Color the map.

Read About It

Name: _____ **Date:** _____

Directions: Read the text. Study the photo. Then, follow the steps.

These are houses. They are over a river. The houses are on stilts. The stilts make them tall. The water does not reach the houses. The houses stay dry.

1. Draw a fish in the river.

2. Circle a stilt.

3. Draw a house on stilts.

© Shell Education

Name: _____ **Date:** _____

Directions: Study the chart. Answer the questions.

Tent	Cabin	Mansion

1. Circle the largest home.

    tent   cabin   mansion

2. How many people can go in the cabin? Circle your answer.

    1   2   3

3. Draw the number of people who can fit in a tent.

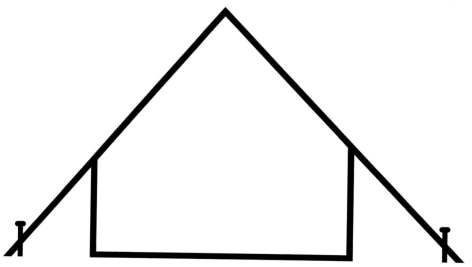

Think About It

Geography and Me

Name: _____ **Date:** _____

Directions: Draw what your house looks like.

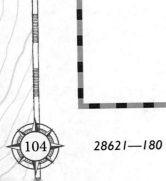

Name: _____ **Date:** _____

Directions: Study the map. Circle the answer to each question.

Reading Maps

1. Does this map show the past or the present?

    past   present

2. What are the houses made of?

    wood   bricks   cement

3. What is used to get places?

    car   bicycle   wagon

Name: _____ **Date:** _____

Directions: What would this map look like today? Draw modern houses and ways of getting around.

Creating Maps

Name: _____ **Date:** _____

Directions: Read the text. Study the photos. Then, circle the answer to each question.

In the past, houses were small. They had only one or two rooms.

Today, houses are bigger. They have many rooms. They have lights.

past

present

1. Houses in the past were _____.

smaller   bigger

2. Houses in the past did not have _____.

doors

windows

lights

Think About It

Name: _____ **Date:** _____

Directions: In the past, buildings had only one floor. Now, buildings can have many floors. Study the drawing. Circle the answer to each question.

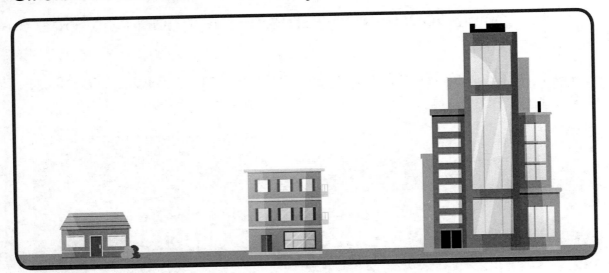

1. Which building holds the most people?

2. Which building only has one floor?

3. Which building is the tallest?

Name: _____ **Date:** _____

Directions: Study the photo. Then, follow the steps.

Past	Present

1. Draw where you live. Draw it next to the photo.

2. Circle two things that are the same in each picture.

3. Put an X on two things that are different in each picture.

Try It! Tell a friend how your house is different from the one in the photo.

Geography and Me

Reading Maps

Name: _____ **Date:** _____

Directions: Study the map. Circle the answer to each question. Then, color the land green.

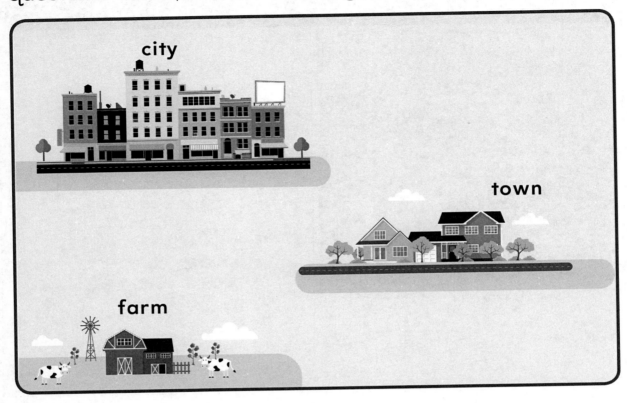

1. Does a town have more buildings than a city?

     yes   no

2. How many homes are on the farm?

     1   5   10

3. Which place has the tallest buildings?

  city   town   farm

Name: _____ **Date:** _____

Directions: Study the map. Follow the steps.

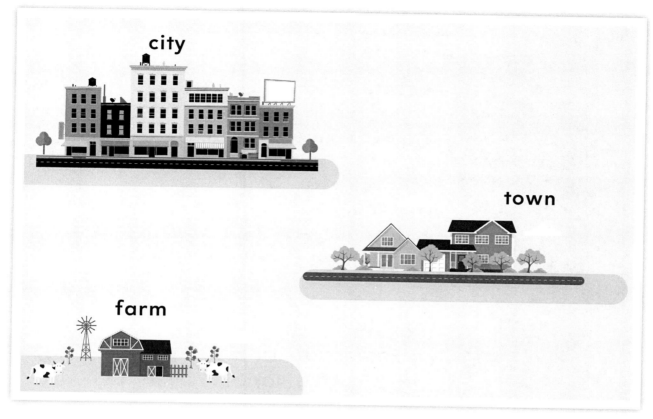

1. Color the tall buildings in the city red.

2. Draw a lake near the town.

3. Circle the farm.

4. Draw things you might see on a farm.

Creating Maps

Read About It

Name: _____ **Date:**_____

Directions: Read the text. Study the photos. Then, circle the answer to each question.

Cities
People in cities live close together. Cities have tall buildings.

Towns
Towns are not as big as cities. People live in houses.

Farms
Farms are not close to cities. They have a lot of land.

1. Farms have a lot of _____.

land

houses

shops

2. Where do people live in tall apartment buildings?

city

town

farm

Name: _____ **Date:** _____

Directions: Study the chart. Circle the answer to each question.

City	Town	Farm

1. How many families live on a farm?

   1   5   10

2. Which place has the most people?

   city   town   farm

3. Which place has the fewest people?

   city   town   farm

Geography and Me

Name: _____ **Date:** _____

Directions: Draw the closest town or city to your house.

Name: _____ **Date:** _____

Directions: Study the map. Circle the answer to each question. Then, color the water blue.

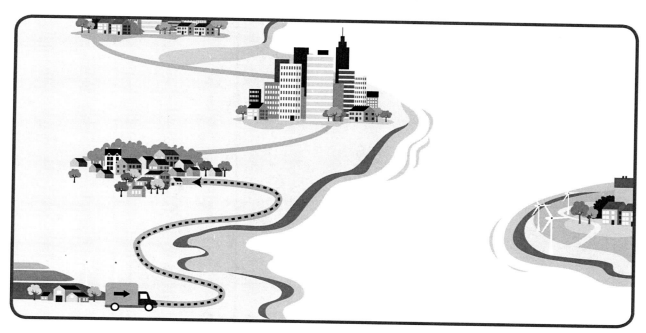

1. This family is moving to a _____.

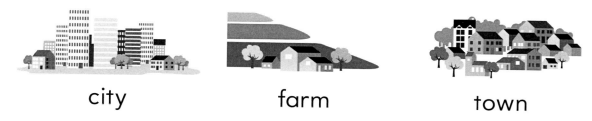

city   farm   town

2. What is the family using to move their belongings?

truck   horse   airplane

3. Their new house is _____ other houses.

near   far from

Name: _____ **Date:** _____

Directions: A family is moving to a city. Draw a line to show where the family will move.

Try It! Tell a friend about the family's move.

Name: _____ **Date:** _____

Directions: Read the text. Study the photo. Then, circle the answer to each question.

> This photo is from a long time ago. It shows people moving. They are called *settlers*. Many left the cities. They moved to open land. Some started farms. Some had ranches.

1. What did the settlers use to move?

    truck   wagon   airplane

2. Some settlers started _____.

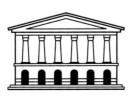      

  museums   ships   farms

Think About It

Name: _____ **Date:** _____

Directions: Study the map. Circle the answer to each question. Then, color the map.

1. Which place might be dangerous?

volcano

town

2. Which place would have the most jobs?

volcano

town

city

Name: _____ **Date:** _____

Directions: Would you like to move someday? Draw a picture of your family moving to a new place.

Reading Maps

Name: _____ **Date:** _____

Directions: Study the pictures. Circle the answer to each question. Then, color the maps.

1. What was the land used for in the past?

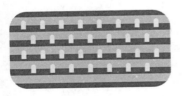   

farming   shopping   playing

2. What is the land used for now?

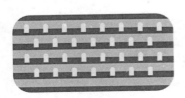   

farming   housing   pasture

Name: _____ **Date:** _____

Directions: The map shows how the land looked in the past. Draw another way the land could be used today.

Read About It

Name: _____ **Date:** _____

Directions: Read the text. Study the photo. Then, circle the answer to each question.

These buildings are restaurants. People eat at them. In the past, they used to be houses. People lived in them. Places change over time.

1. What does the picture show? Circle your answer.

park

train

restaurants

2. How were the buildings used in the past? Circle your answer.

houses

restaurants

libraries

Name: _____   **Date:** _____

Directions: Study the pictures. Answer the questions.

1. Which part of the land is the same in both pictures?
    Color it green.

2. Find the hills on the map from the past. What are
    these hills now? Circle your answer.

trees   city   crops

3. Draw the things that are different in both pictures.

Think About It

Name: _____ **Date:** _____

Directions: Imagine this is your backyard. How would you change the land? Draw your answer.

Name: _____ **Date:** _____

Directions: Study the map. Circle the answer to each question. Then, trace the roads in black.

1. What is near the food market?

café

lake

track

2. What is behind the school?

fountain

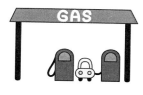

gas station

track

Name: _____ Date: _____

Directions: Draw lines to show where each place is on the map. Then, color the map.

 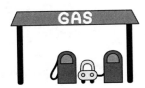 

Name: _____ **Date:** _____

Directions: Read the text. Study the photo. Then, circle the answer to each question.

A community is a place where people live. It is a place where people work. The people help each other. Teachers help kids learn. Farmers grow food. Police keep people safe. Doctors help sick people.

1. Who grows food in a community?

doctor   teacher   farmer

2. Who helps sick people in a community?

doctor   police   farmer

3. Tell a friend about your community.

Think About It

Name: _____ **Date:** _____

Directions: The chart shows how many helpers work in a community. Study the chart. Circle the answer to each question.

Community Helper		Number
police officer		🧍🧍🧍🧍🧍
dentist		🧍
doctor		🧍🧍🧍🧍
firefighter		🧍🧍🧍

1. How many police officers work in this community?

   1   3   5

2. Which job is done by three people?

   firefighter   doctor   dentist

Name: _____ **Date:** _____

Directions: What kind of people live and work in your community? Color and label the people to show who they are.

_____

- - - - - - - - - - - - - -

_____

_____

- - - - - - - - - - - - - -

_____

_____

- - - - - - - - - - - - - -

_____

_____

- - - - - - - - - - - - - -

_____

© Shell Education

Geography and Me

Reading Maps

Name: _____ **Date:** _____

Directions: Study the map. Circle the answer to each question.

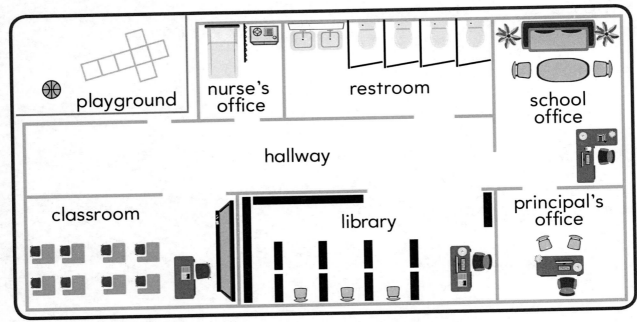

1. Which room is near the restroom?

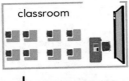

classroom

nurse's office

principal's office

2. What is to the left of the nurse's office?

principal's office

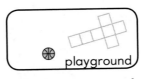

playground

restroom

3. Is the classroom to the left or the right of the library?

left   right

Name: _____ **Date:** _____

Directions: Study the map. Follow the steps.

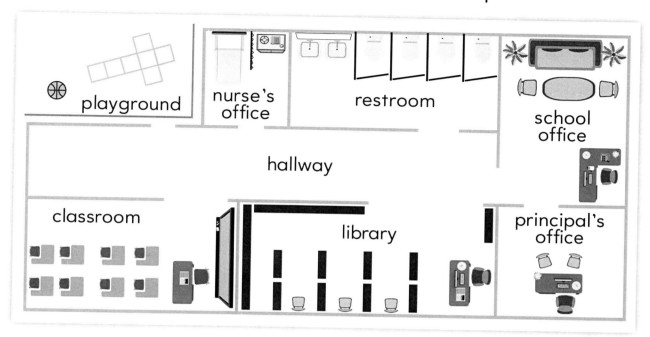

1. Circle the restroom on the map.

2. Put an X on the principal's office.

3. Draw a line from the playground to the nurse's office.

4. Draw your school.

Creating Maps

Read About It

Name: _____ Date: _____

Directions: Read the text. Study the photo. Then, circle the answer to each question.

These are both schools. One school is in the United States. One is in Africa. Students go to both schools to learn. They both have one room. They both have teachers. The schools are different. How many differences can you spot?

African school

American school

1. Which school is made from sticks tied together?

2. What is the roof of the African school made of?

wood   grass   bricks

Name: _____ **Date:** _____

Directions: Read the schedule. Circle the answer to each question.

Time	Activity		Location
9:00	classroom activities		classroom
11:00	physical education (P.E.)		playground
12:00	lunch		cafeteria
1:00	reading		library

1. What does Ms. Robin's class do first?

classroom activities   P.E.   lunch   reading

2. Where does the class go after lunch?

classroom   playground   cafeteria   library

Think About It

Geography and Me

Name: _____ **Date:** _____

Directions: Draw a picture that shows your school community.

Name: _____ **Date:** _____

Directions: Study the map. Draw lines to show where each person goes during the day. Then, color the buildings.

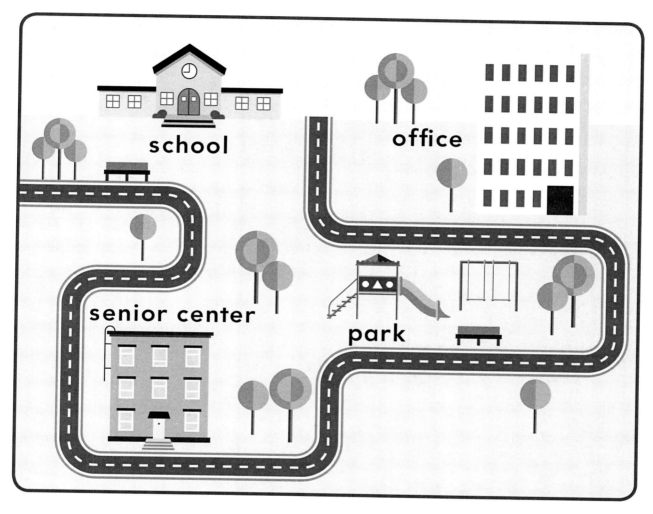

Creating Maps

Name: _____ **Date:** _____

Directions: Draw each person in your family. Show what they each do during the day.

Name: _____ **Date:** _____

Directions: Read the text. Study the photo. Then, follow the steps.

There are many different types of people. Some people are young. Some are old. Some people have light skin. Some have dark skin. We are all different. But we are also the same. We are all people.

1. Draw a circle around a person with dark hair.

2. Draw a triangle on a person wearing stripes.

3. Draw a box around a person with braids.

4. Draw yourself in the box.

Name: _____ **Date:** _____

Directions: The graph shows which students wore blue and which wore red. Study the graph. Answer the questions.

Mr. Chan's Kindergarten Class

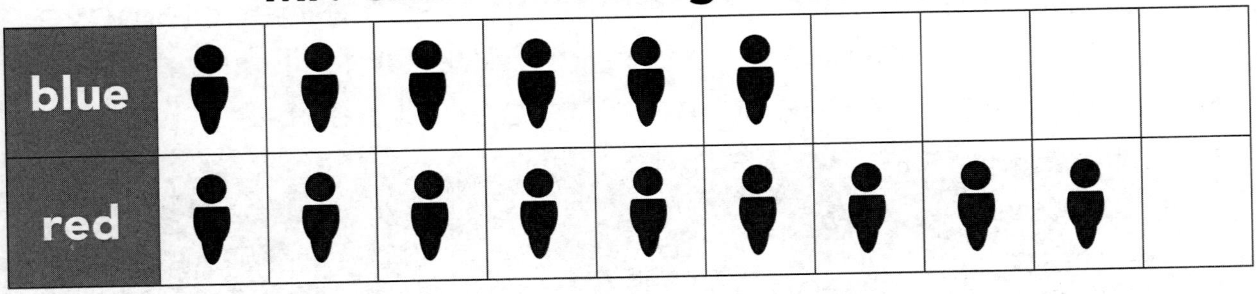

blue	👤	👤	👤	👤	👤	👤				
red	👤	👤	👤	👤	👤	👤	👤	👤	👤	

1. Are there more students who wore blue or red? Circle your answer.

         blue   red

2. How many students wore blue? Circle your answer.

     6   8   10

3. A new boy just joined the class. He was wearing red. Draw a person in the correct column.

4. Draw Mr. Chan's class.

Name: _____ **Date:** _____

Directions: Count the number of students wearing blue and red in your class. Draw symbols to complete the graphs.

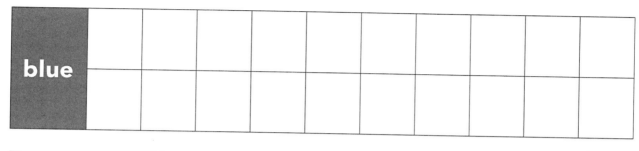

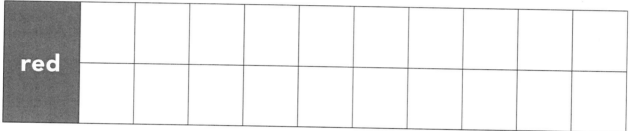

1. Are there more students wearing blue or red in your class? Circle your answer.

    blue   red

2. Draw yourself in your classroom.

Reading Maps

Name: _____ **Date:** _____

Directions: Study the map. Circle the answer to each question. Then, color the buildings.

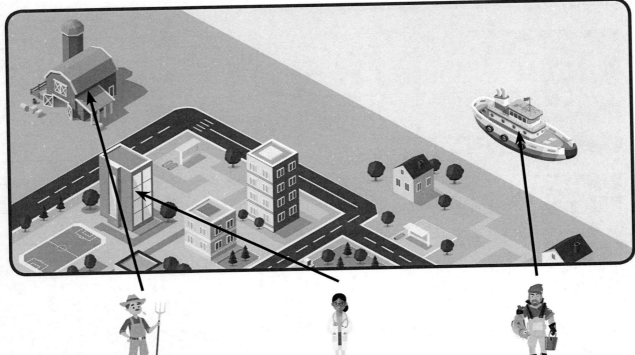

farmer   doctor   fisher

1. Who works on a ship?

farmer   doctor   fisher

2. Where does the farmer work?

farm   town   ship

Name: _____ **Date:** _____

Directions: Draw a line from each person to the place where he or she works. Then, color the buildings.

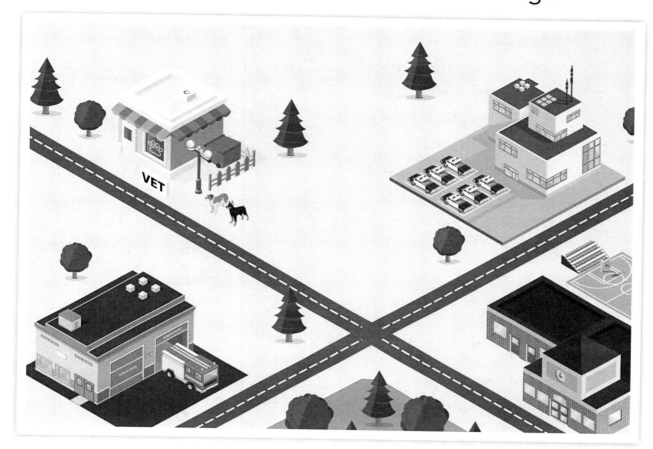

vet

police

firefighter

teacher

Read About It

Name: _____ **Date:** _____

Directions: Read the text. Study the photo. Then, circle the answer to each question.

Where do you want to work? This man works in the ocean. He is a scientist. He studies animals. He learns how to help them. The scientist lives near the ocean. He likes being in the water.

1. Where does the scientist work?

desert   ocean   house

2. What does he study?

ocean animals   insects   jungle animals

3. Could he do his job if he did not live near the water?

yes   no

Name: _____ **Date:** _____

Directions: Study the chart. Circle the answer to each question.

Place	Fire Trucks	Firefighters
city		
town		

1. The city has _____ fire trucks.

    1   3   5   7

2. How many firefighters work in the town?

    1   3   5   7

3. Which place has more firefighters?

         city   town

4. Why might a big city need more firefighters? Tell a friend.

Geography and Me

Name: _____ **Date:** _____

Directions: Draw a picture of yourself doing a job you would like. Show where you will do it. Then, complete the sentences.

_____

I want to be a _____ .

I will do this job in _____ .

Name: _____ **Date:** _____

Directions: Goods are things you buy. Services are things people do for others. Study the picture. Circle the answer to each question.

1. What is one good being sold on this street?

flowers

clothes

books

2. Which place is providing a service?

hospital

market

book store

3. Food is a _____.

good   service

Creating Maps

Name: _____ **Date:** _____

Directions: Study the picture. Follow the steps.

1. Circle a place that provides a service.

2. Draw a store that sells a good in the picture.

3. Draw goods that you use.

© Shell Education

Name: _____ **Date:** _____

Directions: Read the text. Study the photo. Then, circle the answer to each question.

Goods are things that you buy. Bakers make cakes. A cake is a good. A service is something a person does for you. A teacher helps you learn. People need goods and services.

1. What job is shown in the picture?

taxi driver   crossing guard   baker

2. Is this a good or a service?

good   service

3. Which person makes goods?

doctor   police officer   baker

Think About It

Name: _____   **Date:** _____

Directions: Study the picture. Circle the answer to each question.

1. What good is produced by the farmer?

food   truck   forks

2. Does the truck driver provide a good or a service?

good   service

3. Who provides a service in a restaurant?

farmer   server   truck driver

Name: _____ **Date:** _____

Directions: Draw places that provide goods and services in your community. Label the places.

Goods

Services

Geography and Me

Reading Maps

Name: _____ **Date:** _____

Directions: Study the map. Circle the answer to each question. Then, color the map.

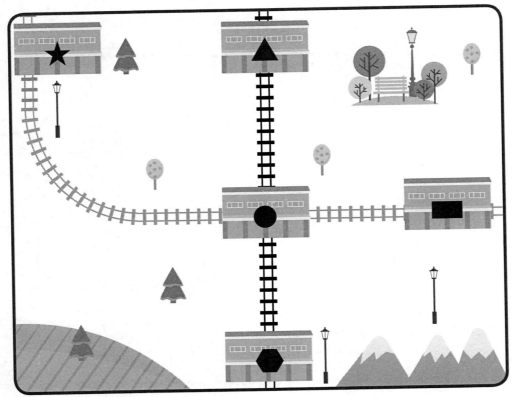

1. Which station is on the light gray track?

2. The is on the _____ line.

3. Where do the two tracks meet?

Name: _____ **Date:** _____

Directions: Draw each bus line on the map. Then, color the map.

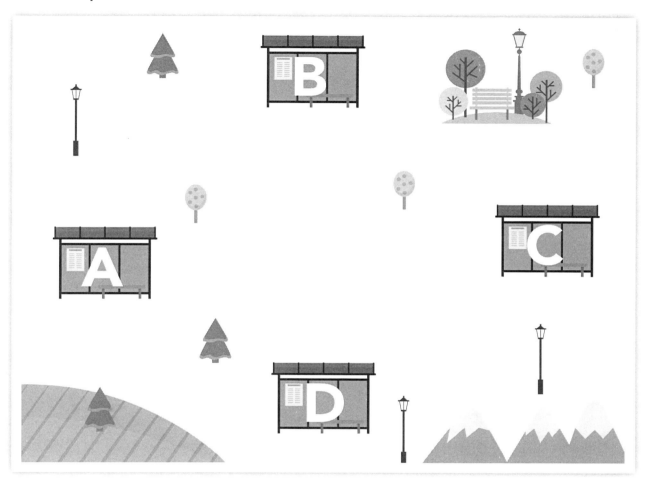

1. The red bus line goes from stop A to stop B.

2. The green bus line goes from stop C to stop D.

3. The blue bus line goes from stop B to stop D.

4. The orange bus line goes from stop A to stop C.

Try It! Tell a friend about the bus lines you drew.

Read About It

Name: _____ **Date:** _____

Directions: Read the text. Study the photo. Then, circle the answer to each question.

In cold places, snow often covers the ground. Cars cannot drive on snow. How do people get around? They use dog sleds! Sleds can go over snow. Dogs pull the sleds. Dog sleds can carry people. They can carry goods.

1. People in which kind of climate use dog sleds?

snowy

sunny

rainy

2. Which of these objects could **not** be carried by a dog sled?

person

groceries

car

3. The dogs run _____ the sled.

     in front of   behind   next to

Name: _____ **Date:** _____

Directions: Study the picture. Circle the answer to each question.

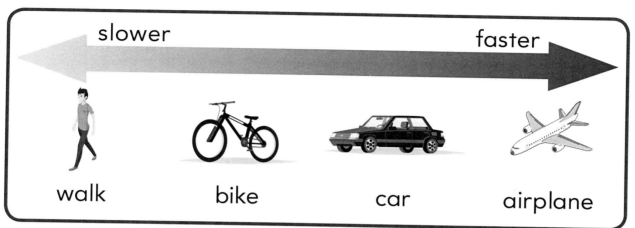

slower ←——————————→ faster

walk   bike   car   airplane

1. Which form of transportation is the fastest?

car   walk   bike   airplane

2. Which is slower, a car or a bike?

car   bike

3. Is a car slower or faster than an airplane?

slower   faster

THINK ABOUT IT

Geography and Me

Name: _____ **Date:** _____

Directions: Study the pictures. Draw a person who would use each type of transportation. Label each person.

Name: _____ **Date:** _____

Directions: Study the map. Circle the answer to each question. Then, color the map.

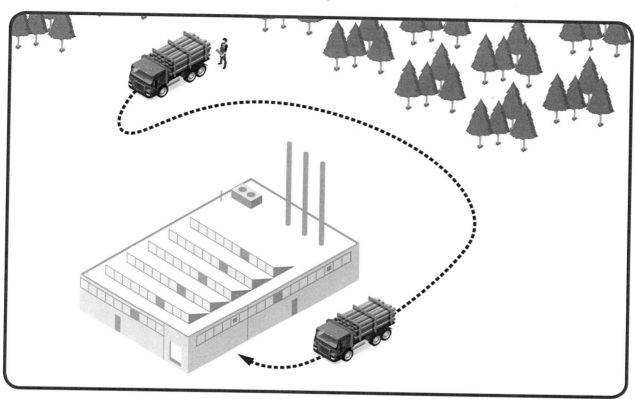

1. How did the logs travel from the forest to the factory?

truck   boat   car

2. Where did the logs end up?

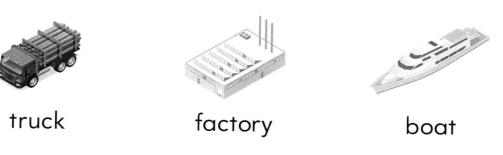

truck   factory   boat

Name: _____ **Date:** _____

Directions: Study the map. Draw a road to show how wood gets from a forest to the paper factory.

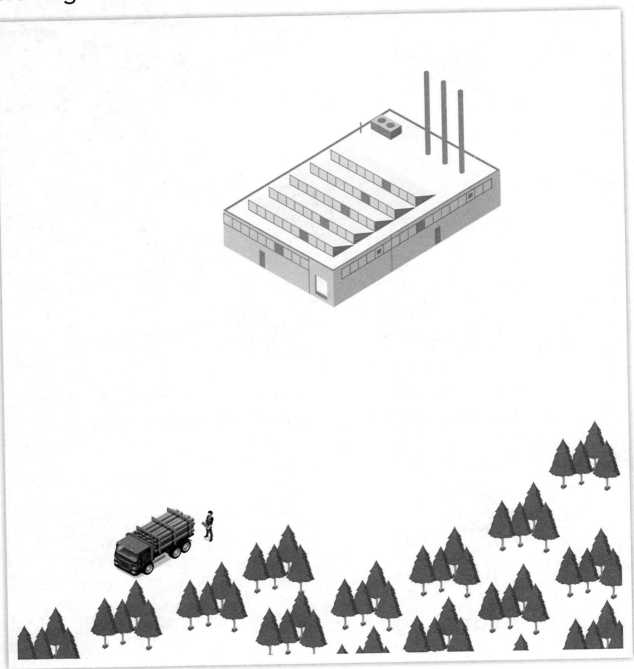

Try It! Tell a friend how the wood gets to the factory.

Name: _____ **Date:**_____

Directions: Read the text, and study the photo. Circle the answer to each question.

Markets sell food. Farmers bring their goods to the market. Many farmers bring their goods in trucks. This picture shows a farmer loading lettuce onto a truck. He will drive the food to the market. The market will sell the food to people.

1. Where is food produced?

market   farm   truck

2. Where is food sold?

truck   farm   market

3. What good is shown in the picture above?

lettuce   eggs   bike

Think About It

Name: _____ **Date:** _____

Directions: Study the graph. Circle the answer to each question.

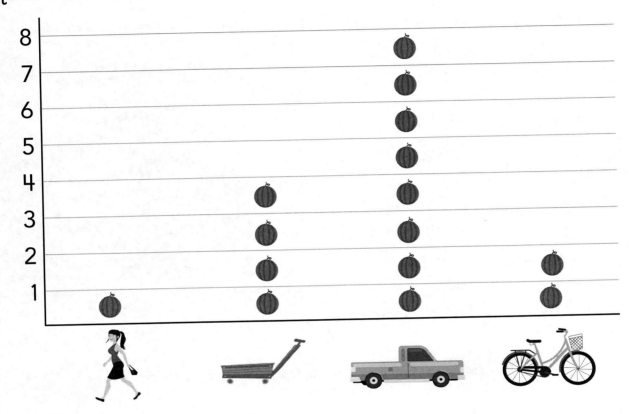

1. Which can carry the most?

person   truck   wagon

2. What should you use to move three melons?

person   wagon   bicycle

Name: _____ **Date:** _____

Directions: Complete the chart. Draw or write your answers.

Goods I Use	Where They Come From

Geography and Me

Name: _____ **Date:** _____

Directions: The picture shows natural resources. Study the map. Circle the answer to each question. Then, color the river blue.

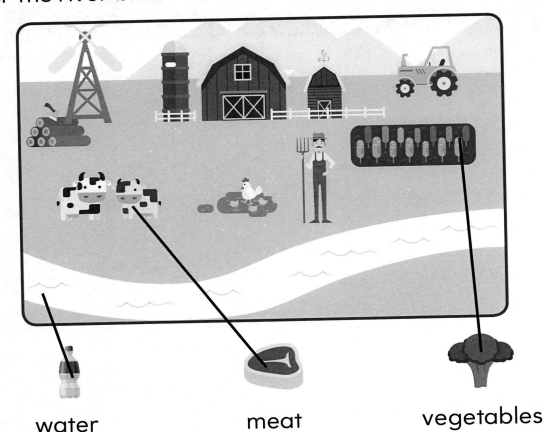

water   meat   vegetables

1. What natural resource do we drink?

water   meat   vegetables

2. Where does meat come from?

garden   mountain   animals

160   28621—180 Days of Geography

© Shell Education

Name: _____ **Date:** _____

Directions: Label the resources shown on the map.

Word Bank

water   meat   vegetables

Read About It

Name: _____ **Date:** _____

Directions: Read the text. Study the photo. Then, circle the answer to each question.

Some goods come from the earth. They are called *natural resources*. Water and soil come from the earth. So do animals. Plants do, too.

These are cotton plants. They come from the earth. Cotton is used to make clothes. Animals eat cotton seeds.

1. Which is not a natural resource?

lettuce   cow   car

2. Which of these things is made from cotton?

shirt   tree   pig

3. Which part of a cotton plant is used as food for animals?

leaf   flower   seed

Name: _____ **Date:** _____

Directions: The chart shows where some natural resources come from. Study the chart. Circle the answer to each question.

Plants	Animals	Earth	Water

1. Cheese comes from _____.

animals   water   earth   plants

2. Which good comes from the earth?

cheese   gems   carrots

3. Which natural resource do we drink?

carrots   gems   water

Name: _____ **Date:** _____

Directions: Draw how you use each kind of natural resource.

Earth

Water

Animals

Plants

Name: _____ **Date:** _____

Directions: Study the map. Circle the answer to each question. Then, color the map.

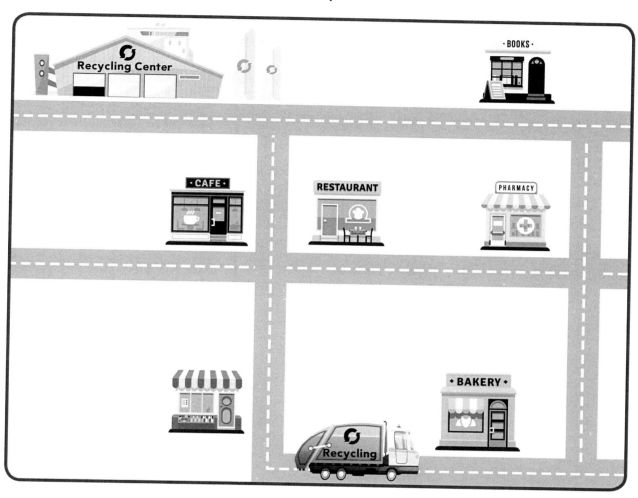

1. What kind of truck is working in this town?

2. Where will the truck go to next?

Name: _____ **Date:** _____

Directions: Study the map. Follow the steps.

Creating Maps

1. Circle the recycling center.

2. Draw a box around the truck.

3. Draw a line to show how the truck gets to the recycling center.

Name: _____   **Date:** _____

Directions: Read the text. Study the photo. Then, circle the answer to each question.

> Paper can be recycled. People use paper. They put it in bins when they are done with it. A truck picks up the bins. The paper goes to a center. It is made into new paper. People buy the new paper. The process starts again.

1. Where should you put paper after it is used?

bin   recycling center   truck

2. Where does the truck take the paper?

recycling center   bin   worker

3. Which is not part of the recycling process?

bin   sewing   truck

Name: _____ **Date:** _____

Directions: Study the chart. Circle the answer to each question.

Resource	How to Save It
water	Turn off the faucet.
gas	Walk or ride your bike.
trees	Recycle paper.

Think About It

1. How can you save water?

Turn off the faucet.   Ride your bike.   Recycle.

2. What can you save by recycling paper?

trees   cows   gas

Name: _____ **Date:** _____

Directions: Draw a picture to show how you can help save natural resources.

Reading Maps

Name: _____ **Date:** _____

Directions: Study the picture. Circle the answer to each question.

1. Where is the mom?

bedroom   dining room   living room

2. Who is in the bathroom?

Mom   Dad   kids

Name: _____ **Date:** _____

Directions: Pretend this is your house. Draw each member of your family in a room. Label each person.

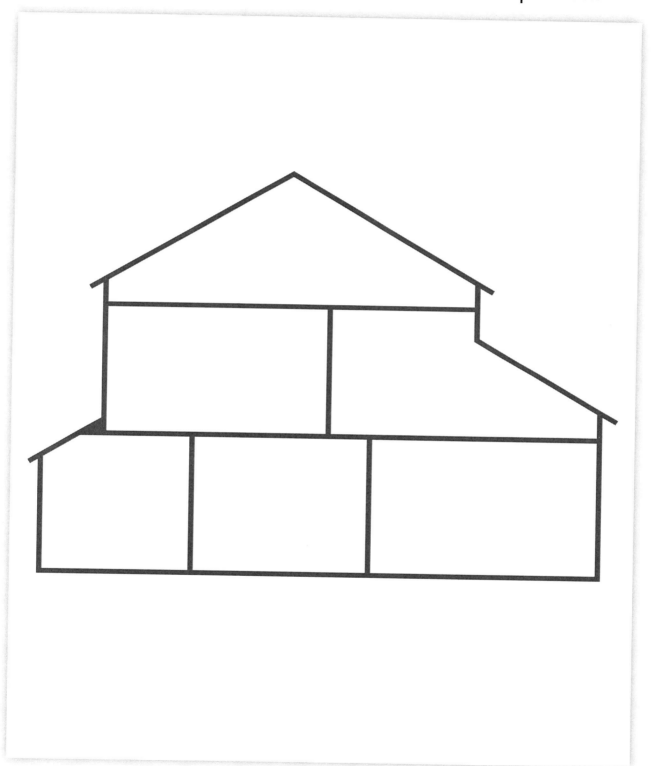

Read About It

Name: _____ **Date:** _____

Directions: Read the text. Study the photo. Then, circle the answer to each question.

This is a big family. They live in India. There are kids. There are adults. There are grandparents. They share a house. They share food. They play and work. They help each other.

1. Circle the youngest person in this family.

2. Where does this family live?

store   house   tent

3. What does this family share?

brush   food   bike

4. Tell a friend about your family.

Name: _____ **Date:** _____

Directions: Study the chart. Circle the answer to each question.

Ken's Family	
mom	1
dad	1
sisters	2
pets	3

1. How many sisters does Ken have?

  1   2   3   4

2. Ken has _____ pets.

  1   2   3   4

3. How many parents does Ken have?

  1   2   3   4

Name: _____ **Date:** _____

Directions: Draw your family. Compare your family to the one in the picture. Write or draw things that are the same and different.

Geography and Me

My Family	The Mendoza Family

Same	Different

Name: _____ **Date:** _____

Directions: Study the map. Follow the steps.

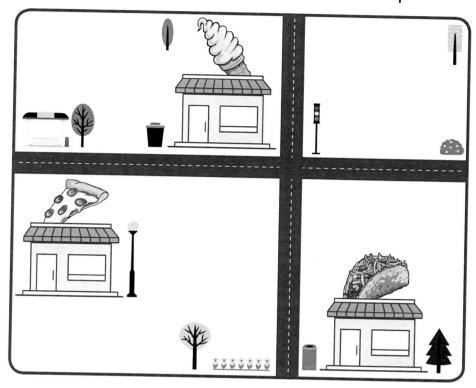

1. Circle the restaurant that serves tacos.

2. Color the ice cream parlor yellow.

3. Put an X on the place where you buy pizza.

4. Draw your favorite restaurant.

Creating Maps

Name: _____ **Date:** _____

Directions: Add two restaurants to the map. Label each one. Draw the type of food each restaurant serves.

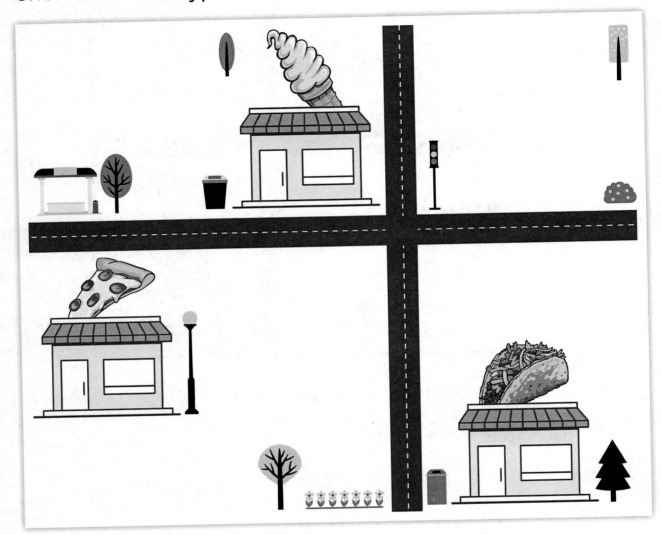

Try It! Tell a friend about the restaurants you drew.

Name: _____ **Date:** _____

Directions: Read the text. Study the photo. Then, circle the answer to each question.

Families eat different food. This is Ethiopian food. It is from Africa. The food is served on special bread. It is soft and flat. You use the bread to scoop up the food. You do not need a fork. You use your hands. It tastes good!

1. How do you eat Ethiopian food?

fork   hands   chopsticks

2. This food comes from _____.

South America   Africa   Australia

3. Tell a friend about a time you tried a new food.

Name: _____ **Date:** _____

Think About It

Directions: Study the graph. Circle the answer to each question.

Favorite Foods in Ms. Martin's Class

1. How many students like hot dogs best?

    2   3   5   8

2. Which food do most students like best?

   spaghetti   burrito   hot dog

Name: _____ **Date:** _____

Directions: Draw a picture of your family eating your favorite food.

Name: _____ **Date:** _____

Directions: Study the map. Circle the answer to each question. Then, color the map.

Reading Maps

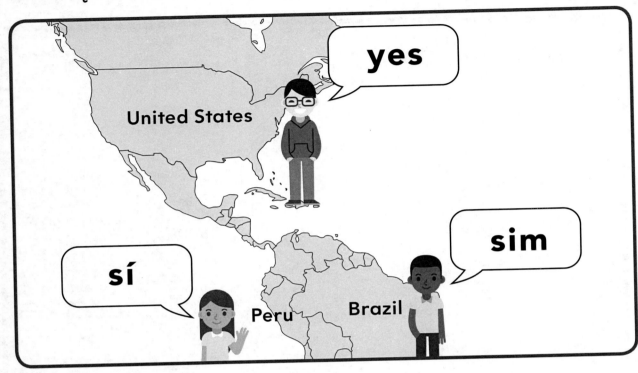

1. How do you say "yes" in Peru?

yes

sí

sim

2. Which child says "sim"?

3. Where is English spoken?

Peru

Brazil

United States

Name: _____

Date: _____

Directions: Complete the map. Write "hello" in each language.

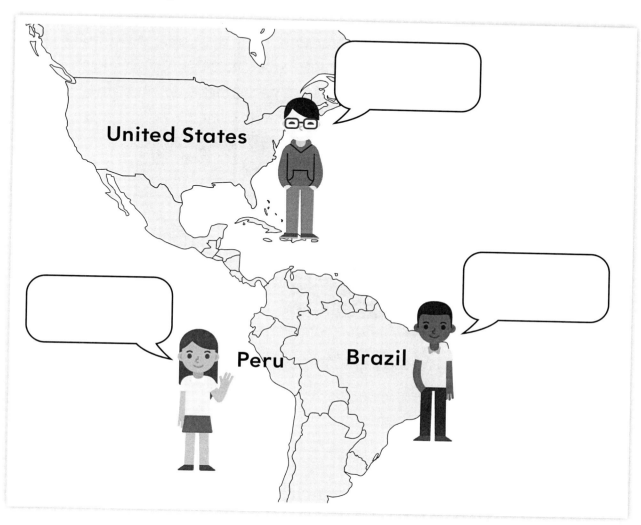

United States—Hello

Peru—Hola

Brazil—Olá

Try It! Tell a friend about the map above.

Creating Maps

Name: _____ **Date:** _____

Read About It

Directions: Read the text. Study the photo. Then, answer the questions.

Languages help us. We use them to talk. We read and write them. Some people speak one language. Some people speak more. Many people speak English. Some speak Spanish. Others speak Hindi. What do you speak?

1. What does the sign say?

_____

- -

_____

2. Each line on the sign is a different language. How many languages are on the sign? Circle your answer.

       1   2   3   4

3. Write the name of one language mentioned in the text.

_____

- -

_____

Name: _____ **Date:** _____

Directions: Five friends all speak English together. At home, they speak different languages. Study the chart. Circle the answer to each question.

Name	Home Language
Caleb	English
Sofia	Russian
Alex	Russian
Oliver	English
Susan	Japanese

1. How many kids speak Russian at home?

1   2   3   4

2. What language does Oliver speak at home?

Japanese   English   Russian

3. How many kids speak Japanese at home?

1   2   3   4

4. Tell a friend what languages the kids speak at home.

Name: _____ **Date:**_____

Directions: Answer the questions.

1. What language do you speak at home?

_____

_ _

_____

2. What languages do your friends speak?

_____

_ _

_____

_____

_ _

_____

3. What language would you like to learn?

_____

_ _

_____

Name: _____ **Date:** _____

Directions: Study the photo. Answer the questions.

Park Clean-Up Day

1. What are these people doing? Circle your answer.

picnicking

cleaning

partying

2. Circle the people who are working together.

3. Tell a friend how you can both work together.

Creating Maps

Name: _____ **Date:** _____

Directions: Draw how people can help each other in a park.

Name: _____ **Date:** _____

Directions: Read the text. Study the photo. Then, circle the answer to each question.

Cooperation means working together. It means helping. It means sharing. Together, people can solve problems.

How can you help? Pick up trash. Help a friend. Be respectful. Share with people. It's easy to help!

1. Cooperation means _____.

winning a prize

being noisy

working together

2. The kids in the picture are helping each other _____.

plant a tree

cook

win a race

Think About It

Name: _____ **Date:** _____

Directions: Study the pictures. Answer the questions.

Is it cooperation?

Yes   No

1. Which is a way to cooperate? Circle your answer.

plant a tree   litter

2. How can you help your parents? Circle your answer.

plant vegetables   fight   step on flowers

3. Tell a friend other ways to cooperate.

Name: _____ **Date:** _____

Directions: How can you work with others to share or care for Earth? Draw your idea. Then, complete the sentence.

I am cooperating by _____ .

Reading Maps

Name: _____ **Date:** _____

Directions: Study the map. Circle the answer to each question.

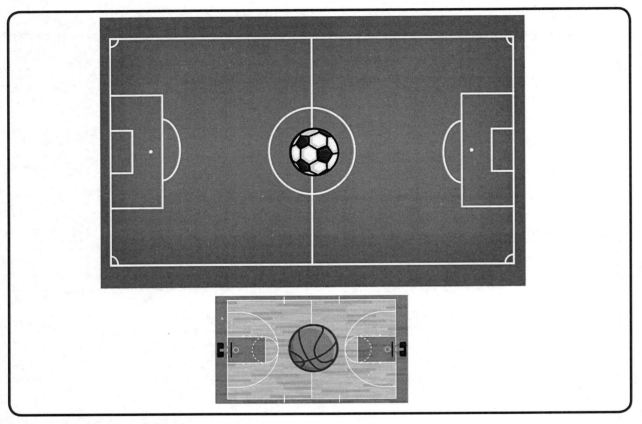

1. Which sport has the most room on the map?

soccer

basketball

2. Which sport would you rather play?

soccer

basketball

Name: _____ **Date:** _____

Directions: Draw a soccer field on one half of the box. Basketball courts are smaller. Draw three basketball courts in the other half.

soccer   basketball

Read About It

Name: _____ **Date:** _____

Directions: Read the text. Study the photo. Then, circle the answer to each question.

People disagree about how to use land. Some people want to build on it. They want more houses. Other people want more open space.

City leaders can help. They have meetings. People share their views. They vote. They decide how to share the land.

1. What might some people want to build on the land?

house

tree

car

2. Who helps decide how to share the land?

kid

worker

city leaders

Name: _____ **Date:** _____

Directions: The chart shows how many kids want each vegetable. Decide where to plant each vegetable. Draw each vegetable in a section of the garden.

Vegetable		Number of Kids
broccoli	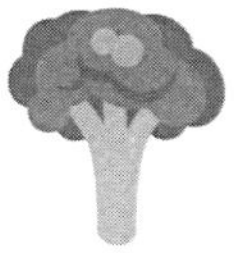	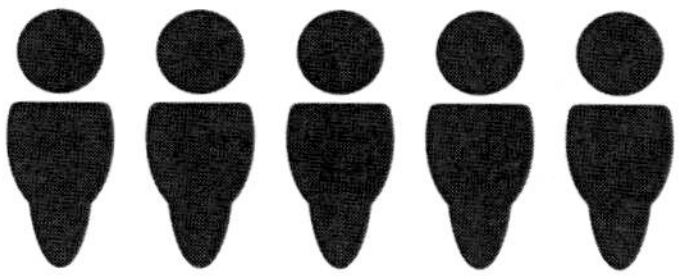 👤👤👤👤👤
carrots		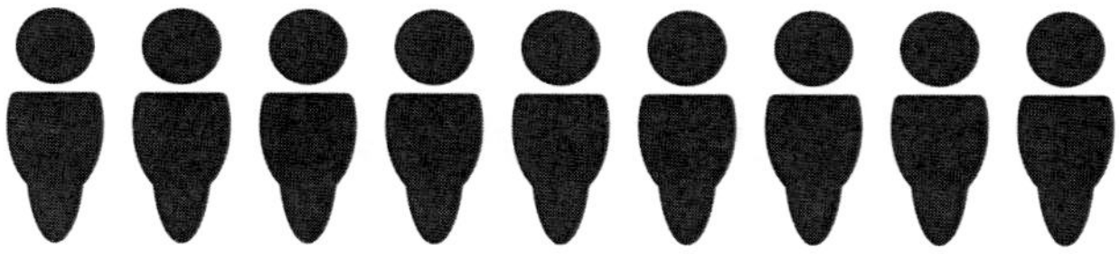 👤👤👤👤👤👤👤👤👤
cucumber		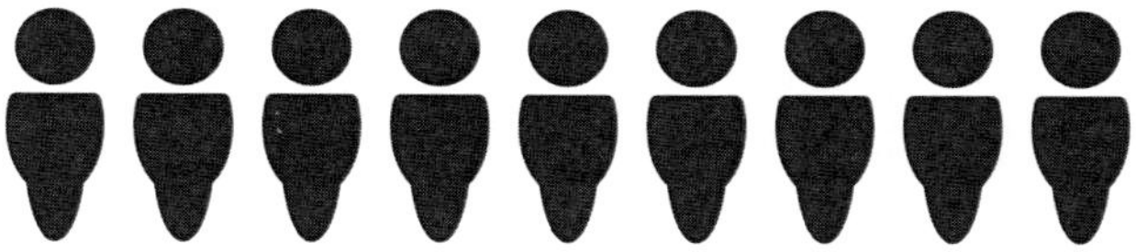 👤👤👤👤👤👤👤👤👤
zucchini	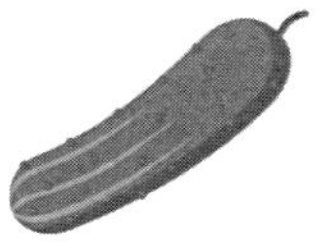	 👤👤

Your Garden

Think About It

Name: _____ **Date:** _____

Directions: You want to play basketball at recess. But other kids are already using the court. What could you do? Draw your answer.

ANSWER KEY

There are many open-ended pages and writing prompts in this book. For those activities, the answers will vary. Answers are only given in this answer key if they are specific.

Week 2 Day 5 (page 24)

1. Students should circle the map.
2. Students should circle the globe.
3. Students should color the second and fifth images blue. Students should put an *X* on the first, third, and fourth images.

Week 3 Day 1 (page 25)

1. over
2. over
3. under

Week 3 Day 3 (page 27)

1. fire
2. pan

Week 3 Day 4 (page 28)

1. toy
2. table
3. over

Week 4 Day 1 (page 30)

1. seesaw
2. monkey bars

Week 4 Day 4 (page 33)

1. 3 squares
2. 2 squares
3. slide

Week 5 Day 1 (page 35)

1. left
2. food
3. left

Week 5 Day 3 (page 37)

1. chair
2. right

Week 6 Day 1 (page 40)

1. gardener
2. plants
3. behind

Week 6 Day 3 (page 42)

1. tree
2. bush

Week 6 Day 4 (page 43)

1. lettuce
2. mint
3. behind

Week 7 Day 1 (page 45)

1. map
2. left
3. skateboard

Week 7 Day 3 (page 47)

1. mat
2. left

Week 8 Day 1 (page 50)

1. Red Street
2. cat
3. 4

Week 8 Day 3 (page 52)

1. Park Street
2. 8

Week 8 Day 4 (page 53)

1. 123
2. Avalon
3. The Long family

Week 9 Day 1 (page 55)

1. The darker color should be circled.
2. The lighter color should be circled.
3. Students should explain how to use a key.

Week 9 Day 3 (page 57)

1. beach
2. Students should draw people swimming.

Week 9 Day 4 (page 58)

1. Students should circle Los Angeles and Boston.
2. Students should draw an *X* on Denver.
3. Los Angeles

Week 10 Day 1 (page 60)

1. mountain
2. plain
3. Students should explain how each landform is different.

ANSWER KEY *(cont.)*

Week 10 day 3 (page 62)

1. A tree should be drawn at the top of a hill.
2. hills

Week 10 Day 4 (page 63)

1. plain
2. mountain

Week 11 Day 1 (page 65)

1. pond
2. ocean
3. river

Week 11 Day 3 (page 67)

1. mountains
2. river

Week 11 Day 4 (page 68)

1. ocean
2. puddle

Week 12 Day 1 (page 70)

1. windy
2. rainy

Week 12 Day 3 (page 72)

1. rainy
2. rainy and snowy

Week 12 Day 4 (page 73)

1. sunny
2. winter
3. yes

Week 13 Day 1 (page 75)

1. square
2. star
3. circle

Week 13 Day 2 (page 76)

1. Students should color the uppermost region purple, the next region blue, the next region green, and the most southern region yellow.

Week 13 Day 3 (page 77)

1. sunny
2. Students should circle the lizard, snake, and bird.
3. Students should draw two different cactus shapes.

Week 13 Day 4 (page 78)

1. Kate
2. 5
3. Manny

Week 14 Day 1 (page 80)

1. trees
2. cacti

Week 14 Day 3 (page 82)

1. grass
2. cold
3. snowy

Week 14 Day 4 (page 83)

1. tundra
2. desert

Week 15 Day 1 (page 85)

1. Students should draw pond animals such as fish, turtles, frogs, ducks, dragonflies, and butterflies.
2. Students should draw pond plants such as trees, bushes, cattails, and lily pads.

Week 15 Day 3 (page 87)

1. bug
2. bird
3. lily pad

Week 15 Day 4 (page 88)

1. leaf
2. bird
3. grasshopper

Week 16 Day 1 (page 90)

1. rainy
2. coat
3. snowy

Week 16 Day 3 (page 92)

1. cold
2. coat
3. Answers may include coats, hats, scarves, gloves, and pants.

ANSWER KEY *(cont.)*

Week 16 Day 4 (page 93)
1. sunglasses
2. feet
3. sunny

Week 17 Day 1 (page 95)
1. hill
2. skate

Week 17 Day 2 (page 96)

kite flying: hill
swimming: lake
hiking: mountains

Week 17 Day 3 (page 97)
1. beach
2. sunny

Week 17 Day 4 (page 98)
1. swim
2. field

Week 18 Day 1 (page 100)
1. igloo
2. apartment

Week 18 Day 4 (page 103)
1. mansion
2. 2
3. 1

Week 19 Day 1 (page 105)
1. past
2. wood
3. horse and wagon

Week 19 Day 3 (page 107)
1. smaller
2. lights

Week 19 Day 4 (page 108)
1. three–story apartment building
2. one–story house
3. three–story apartment building

Week 20 Day 1 (page 110)
1. no
2. 1
3. city

Week 20 Day 3 (page 112)
1. land
2. city

Week 20 Day 4 (page 113)
1. 1
2. city
3. farm

Week 21 Day 1 (page 115)
1. town
2. truck
3. near

Week 21 Day 3 (page 117)
1. wagon
2. farms

Week 21 Day 4 (page 118)
1. volcano
2. city

Week 22 Day 1 (page 120)
1. farming
2. housing

Week 22 Day 3 (page 122)
1. restaurants
2. houses

Week 22 Day 4 (page 123)
1. Students should color the hill, grass, two trees, and bushes on both maps green.
2. crops
3. Answers may include the crops, cow, fence, road, barn, and deer.

Week 23 Day 1 (page 125)
1. cafe
2. track

ANSWER KEY *(cont.)*

Week 23 Day 2 (page 126)

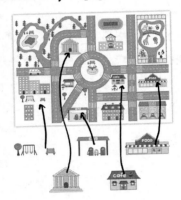

Week 23 Day 3 (page 127)

1. farmer
2. doctor
3. Students should describe their communities.

Week 23 Day 4 (page 128)

1. 5
2. firefighter

Week 24 Day 1 (page 130)

1. nurse's office
2. playground
3. left

Week 24 Day 3 (page 132)

1. African school
2. grass

Week 24 Day 4 (page 133)

1. classroom activities
2. library

Week 25 Day 1 (page 135)

boy: school
man: office
toddler: park
elderly woman: senior center

Week 25 Day 4 (page 138)

1. red
2. 6
3. Students should draw another figure in the "red" column of the graph.
4. Students should draw 10 children wearing red and six children wearing blue.

Week 26 Day 1 (page 140)

1. fisher
2. farm

Week 26 Day 2 (page 141)

Week 26 Day 3 (page 142)

1. ocean
2. ocean animals
3. no

Week 26 Day 4 (page 143)

1. 5
2. 3
3. city
4. Answers should include that there are more people or buildings.

Week 27 Day 1 (page 145)

1. flowers
2. hospital
3. good

Week 27 Day 3 (page 147)

1. crossing guard
2. service
3. baker

Week 27 Day 4 (page 148)

1. food
2. service
3. server

Week 28 Day 1 (page 150)

1. star
2. gray
3. circle

ANSWER KEY *(cont.)*

Week 28 Day 3 (page 152)
1. snowy
2. car
3. in front of

Week 28 Day 4 (page 153)
1. airplane
2. bike
3. slower

Week 29 Day 1 (page 155)
1. truck
2. factory

Week 29 Day 3 (page 157)
1. farm
2. market
3. lettuce

Week 29 Day 4 (page 158)
1. truck
2. wagon

Week 30 Day 1 (page 160)
1. water
2. animals

Week 30 Day 2 (page 161)

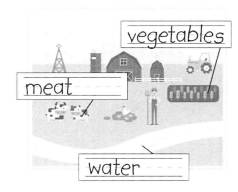

Week 30 Day 3 (page 162)
1. car
2. t-shirt
3. seed

Week 30 Day 4 (page 163)
1. animals
2. gems
3. water

Week 31 Day 1 (page 165)
1. recycling truck
2. bakery

Week 31 Day 3 (page 167)
1. bin
2. recycling center
3. sewing

Week 31 Day 4 (page 168)
1. turn off the faucet
2. trees

Week 32 Day 1 (page 170)
1. dining room
2. Dad

Week 32 Day 3 (page 172)
1. Students should circle the youngest member of the family.
2. house
3. food
4. Students should discuss their families.

Week 32 Day 4 (page 173)
1. 2
2. 3
3. 2

Week 33 Day 3 (page 177)
1. hands
2. Africa

Week 33 Day 4 (page 178)
1. 5
2. burrito

Week 34 Day 1 (page 180)
1. sí
2. Brazilian child
3. United States

ANSWER KEY *(cont.)*

Week 34 Day 2 (page 181)

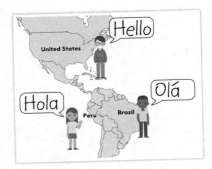

Week 34 Day 3 (page 182)

1. Temple St.
2. 2
3. Students should write either English, Spanish, or Hindi.

Week 34 Day 4 (page 183)

1. 2
2. English
3. 1
4. English, Russian, and Japanese

Week 35 Day 1 (page 185)

1. cleaning
2. Students should circle people that are working together in the picture.

Week 35 Day 3 (page 187)

1. working together
2. plant a tree

Week 35 Day 4 (page 188)

1. plant a tree
2. plant vegetables
3. Answers may include cleaning up, playing fair, and being nice.

Week 36 Day 1 (page 190)

1. soccer

Week 36 Day 3 (page 192)

1. houses
2. city leaders

WORLD MAP

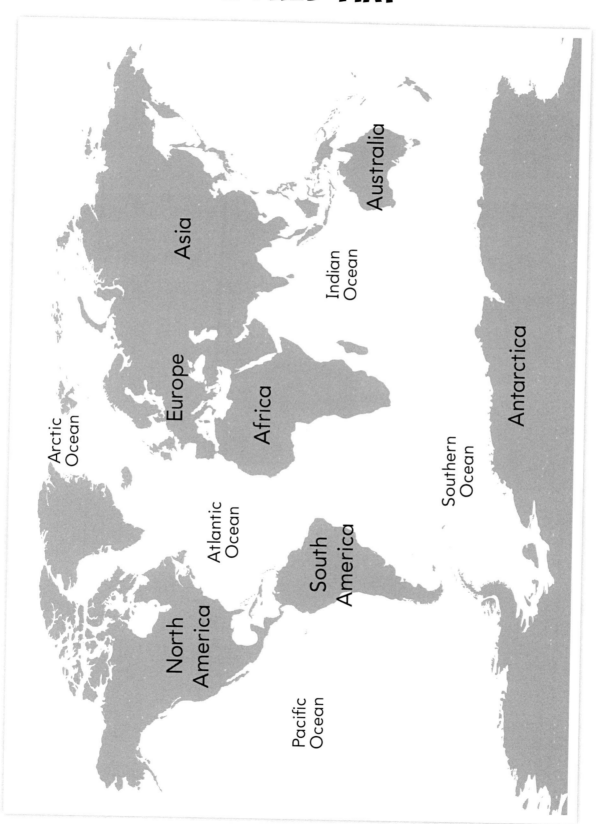

MAP SKILLS RUBRIC
DAYS 1 AND 2

Directions: Evaluate students' activity sheets from the first two weeks of instruction. Every five weeks after that, complete this rubric for students' Days 1 and 2 activity sheets. Only one rubric is needed per student. Their work over the five weeks can be evaluated together. Evaluate their work in each category by writing a score in each row. Then, add up their scores, and write the total on the line. Students may earn up to 5 points in each row and up to 15 points total.

Skill	5	3	1	Score
Identifying Map Features	Identifies and uses all features on maps, including keys and symbols.	Identifies and uses most features on maps.	Does not identify and use features on maps.	
Using Directional Words	Uses directional words to accurately locate places all or nearly all the time.	Uses directional words to accurately locate places most of the time.	Does not use directional words to accurately locate places.	
Interpreting Maps	Accurately interprets maps to answer questions all or nearly all the time.	Accurately interprets maps to answer questions most of the time.	Does not accurately interpret maps to answer questions.	

Total Points: _____

APPLYING INFORMATION
AND DATA RUBRIC
DAYS 3 AND 4

Directions: Complete this rubric every five weeks to evaluate students' Day 3 and Day 4 activity sheets. Only one rubric is needed per student. Their work over the five weeks can be evaluated together. Evaluate their work in each category by writing a score in each row. Then, add up their scores, and write the total on the line. Students may earn up to 5 points in each row and up to 15 points total. **Note:** Weeks 1 and 2 are map skills only and will not be evaluated here.

Skill	5	3	1	Score
Interpreting Texts	Correctly interprets texts to answer questions all or nearly all the time.	Correctly interprets texts to answer questions most of the time.	Does not correctly interpret texts to answer questions.	
Interpreting Data	Correctly interprets data to answer questions all or nearly all the time.	Correctly interprets data to answer questions most of the time.	Does not correctly interpret data to answer questions.	
Applying Information	Applies new information and data to known information about places all or nearly all the time.	Applies new information and data to known information about places most of the time.	Does not apply new information and data to known information about places.	

Total Points: _____

Name: _____ **Date:** _____

MAKING CONNECTIONS RUBRIC
DAY 5

Directions: Complete this rubric every five weeks to evaluate students' Day 5 activity sheets. Only one rubric is needed per student. Their work over the five weeks can be evaluated together. Evaluate their work in each category by writing a score in each row. Then, add up their scores, and write the total on the line. Students may earn up to 5 points in each row and up to 15 points total. **Note:** Weeks 1 and 2 are map skills only and will not be evaluated here.

Skill	5	3	1	Score
Comparing One's Community	Makes meaningful comparisons of one's own home or community to others all or nearly all the time.	Makes meaningful comparisons of one's own home or community to others most of the time.	Does not make meaningful comparisons of one's own home or community to others.	
Comparing One's Life	Makes meaningful comparisons of one's daily life to those in other locations all or nearly all the time.	Makes meaningful comparisons of one's daily life to those in other locations most of the time.	Does not make meaningful comparisons of one's daily life to those in other locations.	
Making Connections	Uses information about other locations to make meaningful connections about life there all or nearly all the time.	Uses information about other locations to make meaningful connections about life there most of the time.	Does not use information about other locations to make meaningful connections about life there.	

Total Points: _____

MAP SKILLS ANALYSIS

Directions: Record each student's rubric scores (page 202) in the appropriate columns. Add the totals, and record the sums in the Total Scores column. Record the average class score in the last row. You can view: (1) which students are not understanding map skills and (2) how students progress throughout the school year.

Student Name	Week 2	Week 7	Week 12	Week 17	Week 22	Week 27	Week 32	Week 36	Total Scores
Average Classroom Score									

APPLYING INFORMATION AND DATA ANALYSIS

Directions: Record each student's rubric scores (page 203) in the appropriate columns. Add the totals, and record the sums in the Total Scores column. Record the average class score in the last row. You can view: (1) which students are not understanding how to analyze information and data and (2) how students progress throughout the school year.

Student Name	Week 7	Week 12	Week 17	Week 22	Week 27	Week 32	Week 36	Total Scores
Average Classroom Score								

MAKING CONNECTIONS ANALYSIS

Directions: Record each student's rubric scores (page 204) in the appropriate columns. Add the totals, and record the sums in the Total Scores column. Record the average class score in the last row. You can view: (1) which students are not understanding how to make connections to geography and (2) how students progress throughout the school year.

Student Name	Week 7	Week 12	Week 17	Week 22	Week 27	Week 32	Week 36	Total Scores
Average Classroom Score								

DIGITAL RESOURCES

To access the digital resources, go to this website and enter the following code: 22738655.
www.teachercreatedmaterials.com/administrators/download-files/

Rubrics

Resource	Filename
Map Skills Rubric	skillsrubric.pdf
Applying Information and Data Rubric	datarubric.pdf
Making Connections Rubric	connectrubric.pdf

Item Analysis Sheets

Resource	Filename
Map Skills Analysis	skillsanalysis.pdf skillsanalysis.docx skillsanalysis.xlsx
Applying Information and Data Analysis	dataanalysis.pdf dataanalysis.docx dataanalysis.xlsx
Making Connections Analysis	connectanalysis.pdf connectanalysis.docx connectanalysis.xlsx

Standards

Resource	Filename
Standards Charts	standards.pdf